LIGHT
REVOLUTIONS

Geoffrey Ernest Stedman

Caption to cover picture Cover Picture the UG2 ring laser at Cashmere New
Zealand with Professor Ulli Schreiber, copyright to the Christchurch Press,
Fairfax NZ.

Caption to frontispiece picture p 1. Lyttelton harbor is visible as a coastal
indentation to an old volcanic crater the lava flows on the north (upward)
flank separate the harbor from the city and the Cashmere Cavern is on this
north flank of these lava flows. Acknowledgements to NASA.

Rev. date: 03/14/2017

To order additional copies of this book, contact:
Xlibris
1-800-455-039
www.xlibris.com.au
Orders@Xlibris.com.au

Contents

1 Preface

Small ring laser gyroscopes were developed in the decades surrounding the 1970's for avionics the principal advance being through advances in mirror quality. Photos of avionic gyros are not readily available since the technology involved is highly sensitive commercial information. One system is illustrated in[1]

Figure 3, an aircraft-type ring-laser gyro

[2] Such devices are the basis of avionic navigation systems and have areas of $(dm)2$ i.e. ~60cmx 60cm as befits airline gyros. as opposed to the areas of up to over 300 m² of the devices discussed here see chapter 7.

We describe in this book not only some amazing properties of light but a particular program of laser gyroscope development, which is aimed to measure local variations in Earth rotation, this has expanding rapidly several major members of a 'family' or 'farm' of earth rotation devices of Figure 3, were built initially at cashmere New Zealand, all of which were much larger than the avionic devices of figure 3. It is the stuff of science that our dreams can be inadequate, certainly mine were. The story I shall tell was completely unexpected and unplanned in my scientific career. I remember as a postgraduate student in the 1960's hearing lasers were described at scientific conferences as "a solution in search of a problem." Today laser's are a multi-billion dollar business. Every supermarket checkout, every CD player and builders' toolkits hold one.

My aim is to work though some concepts to do with the origins of a major experimental project till 2001 and to give an overview of its history from1970 to the present. The Canterbury ring laser project had the ambition of gaining new results on earth rotation by optical means. Its origins are explained as a case study of several of the issues outlined above. For example It depends on the Sagnac effect (chapter 2) to measure the rate of rotation. This has sparked un-necessary misguided debates over light travel partly because of the tendency for confusion and inconsistent notions of simultaneity in relativity this topic is explored in Chapter 4. (It is important that the reader gains some general appreciation of relativity and quantum mechanics two 19th Century revolutions in physics) brief introductions are included here Chapters 5 6.

This book was originally conceived on the basis of an adult education course I gave at the University of Canterbury in 2000. The course discussed theories of light down the centuries and a full description on the ring laser project, what follows is a compressed account the but also other topics related to light and electromagnetism, one of these is the topic of Chapter 10 namely a discussion of hidden momentum, another briefly mentioned in chapter 6 is a lecture demonstration of a physics experiment proving that we cannot know as much as we would like to about the natural world, a full description is in[3].

This illustrates one of the odd-ball topics I touched on during this course. In particular this requires a discussion of topics like the vector potential of electromagnetic induction and momentum conservation in it. These also are off shoots of the theory of light and some aspects of this are poorly understood even by professional physicists and I think they deserve an airing. Even in a book mainly devoted mainly to our experimental project on ring lasers as rotation sensors. These are based on two counter-revolving beams of laser light. an understanding of the operation of an optical gyroscope (is given in Chapter 2 on the Sagnac effect).

Light revolutions

In Grecian days it was imagined that light was composed of particles that were each copies of the object being looked at.

The pioneering work of Isaac Newton was also based on the concept that light was a beam of particles which obeyed his laws of mechanics, this model had some successes which in retrospect were very surprising (however Newton's approach also had some major problems (e.g. such interference effects known as Newton's Rings seen when a convex glass is rested on a flat glass plate, to accommodate which Newton devised some extensions to his picture of light inventing the idea of fits of reflection one which is today unnecessary.)

But in the 1800's conclusive evidence was found that light must also be recognized as a wave, as sound had been. This in turn raised fundamental questions. It was known that sound needed a medium like air to travel in, what was the equivalent for light. A medium dubbed the 'ether' was postulated to carry the wave motion of light. Efforts to detect this medium were not successful and eventually the concept of the ether was abandoned in science.

In the 20th century big advances in Physics were made in relativity and the quantum theories both of which have now received overwhelming experimental confirmation. The particle idea (though not the Greek's idea of copies) was found to have much truth and particles of light are called photons. This had been vindicated by the huge advances in electromagnetism and optical studies in the 19th Century by Maxwell who following the astonishing experimental insight of Faraday showed that one can understand the properties of light particularly the wave properties in terms of oscillations in electromagnetic fields, the electric and magnetic fields in space. Although this fact is well known and is described in detail in many books there are some aspects of electromagnetism and optics which are very poorly known even today so I will detour into a popular if incomplete account of some of these at various stages. The conflict between models raised many major puzzles in Physics whose full understanding has taken the 20th Century advances to elucidate.

Enormous effort was spent by many scientists on trying to resolve the apparent conflict between Wave and Particle models of light but the outcome was that both proved essential. That the character of light was finally understood to be neither wave nor particle on its own but more profound than both. The full understanding of this required grappling with a number of major mysteries whose full solution has required well over a century of dedicated effort by scientists. I cannot review all of this here but outline some relevant considerations as far as is possible, in a work of this size. Nothing rivals a full physics text book account and I do not offer that. I aim to illustrate though aspects of the physical problems relevant to our main application the ring laser project so that a general reader may understand something of the background reasons for undertaking this scientific project.

It was one of the great ironies of scientific history that the very experiments devised to confirm Maxwell's wave theory of light and the generation of radio waves were later recognized as providing clear evidence of the particle nature of light. This was because when light from an electric spark employed by Hertz was allowed to fall on the receiver electrodes, the sparks demonstrating the reception of radio waves came more copiously, because. The ultraviolet rays in the transmitter sparks eject of electrons from the receiver electrodes by the photoelectric effect a particularly quantum mechanical story I have no room to discuss here.

This may help an appreciation as to why all relevant matters one must have some basic understanding of quantum mechanics and relativity. A reader skeptical of science should understand that the basic ideas I outline here are now no longer speculative but fully confirmed by exhaustive experiments some of the early ones as regards ring lasers are briefly reviewed elsewhere[4]. But the light revolutions that will take most of our attention here. Are not the revolutions in our understanding of light itself but the counter rotation of beams of laser light in optical gyroscopes which are now standard in applied science including avionics.

I review here a family of ring-laser devices in New Zealand and now other countries especially Germany which are using ring lasers to measure the effects of earth rotation fluctuations by interfering counter-rotating light beams.

This project has been a major research topic at my institution the Department of Physics University of Canterbury Christchurch New Zealand since 1998. It was a new development at the University of Canterbury conducted in association with scientific groups at State University of Oklahoma Stillwater USA's and later with the TUM: Technical University of Munich, through their Fundamental Research Station Wetzell in the foothills of Bavaria Germany. This institution and our program was also supported by the BKG Bundesamt für Cartography and Geodäsie in Frankfurt also the Technical University of Munich TUM. As a result of this joint program it has proved possible to measure a number of geophysical effects in a novel way. The discovered novel effects outlined here include

- the local rotational effects of ocean and earth tides

- Earthquakes waves both local and remote which generate rotational motions.

- Also the effect of the Moon's gravity on the polar axis of the Earth, The moon's gravitational pull on the in-homogeneities in the Earth creates a diurnal wobble of the axis of rotation of the Earth whose magnitude is 60 cm at the earth's pole,

- also the Chandler Wobble another of the earth natural motions.

- via their effect on the local rotation of the earth was measured by one of the ring lasers discussed here

This was made possible by building a family of ring lasers to measure earth rotation via the relative frequency of co- and counter-rotating laser's beams. The New Zealand devices started of at 1 m square C-I but grew to UG2 the size of our cavern laboratory at Cashmere Christchurch. The German machine, G is about 4 m square. Similar projects have been developed at the small end of the scale in size but at Piñon Flats California for earth quake studies and Pisa Italy for related fundamental studies. The main purpose of this book is to explain the history of these projects at Canterbury and some of the background scientific thinking which stimulated them, the more technical details are generally omitted and no full attempt is made to explain any of these various effects in detail.

The story started for me with an appreciation of the subtleties of the definition of synchronization in relativity this became an interest for mine as a student. And then as a lecturer later with two of my students Ron Anderson and Kumar Vetheraniam (chapter 4). My background was that of a theoretical solid state physicist but not particularly either experimental or laser physics Chapter 4 may help to explain why someone like me should became engaged in managing a new experimental project requiring a new laboratory complex while maintaining research in theoretical physics.

So the bulk of this work is to do with ring lasers. However I include a summary of some of the more exotic concepts (chapter 10) that helped formed a backbone to that adult education course. My aim will be to give a plain mans account accessible to anyone with an interest in science and who has perhaps wondered what really has been going on at Cashmere. Chapter 3 outlines the Cashmere cavern laboratory and Chapter 7 the 'farm' of devices built there also some collaboration built devices.

Of course all scientific work is done within a historical framework. Larger laser gyroscopes go back to the earliest days of lasers. Macek and Davis historic experiment on a rotating 1-meter square Helium-Neon ring laser was in 1962. In the pre-laser era of 1870's one of the closest analogy to today's ring lasers is

Lodge's whirling machine.

(a), (b):

Figures 4, (a),(b)[5] : Lodges whirling machine[6] which aimed to set the ether in motion, ether being the imaginary material which in the eyes of many British Physicists carried light waves. The continental physicist Henry Poincaire on the other hand wrote of the impotence of the ether and said that even if such experiments were 100 times more sensitive the results would still be negative. The ether slowly disappeared from science. The wooden case at the top contained an optical interferometer and some metal blades whirled at high speed 2800 rpm by the electric motor below so as to stir up the ether and distinguish the speeds of light in opposite directions. No such effect was not found. ('Presumably it was counted advisable to keep ones head below the plane of the disks) to avoid them being sheared should the disks fracture.' In another experiment the disks were replaced by an iron spheroid 3' across ½' thick weighing ½ ton. The ether was still un-obligingly obstinate in revealing itself. And when the machine bucked and smoked at 300 rpm. The scientists called it quits. The foundations of the device were secured to the Liverpool sandstone. Lodge anticipated the Sagnac effect (Chapter 2)which recently passed its 100th anniversary and hoped to detect the the rotation of the earth, in the lower photo Oliver Lodge is on the left watching his assistant Benjamin Davis adjust the machine, on the right is George Holt who paid for the apparatus. Although Larmour tweaked Lodge that he would have to stop the earth to check any effect.[7] In fact all he would need as explained mentioned in the chapter 7 was a good sound system to convert the interferometer fringes he saw to an audio signal. Sadly sound systems were unknown in those days.

General references are given and footnotes are at the end of each Chapter, some papers often referred to more widely are noted as GEN:A-I and are listed at the end of each chapter 11.

Acknowledgements Unaccredited photos are by mostly by University staff. It is impossible to give an adequate list of acknowledgements any fair list would be extraordinarily long, given the debt I owe to so many who have helped me in my scientific life. I would thank in particular my wife Rachel and brother David Stedman, Clive Rowe and Dr Bob Hurst for help in editing and Dr Rod Syme for discussions over hidden momentum in 2016 revision Dr D. J. Newman & R.A.Stedman.

Endnotes

[1]   http://en.wikipedia.org/wiki/Ring_laser_gyroscope

[2]   I have described my own lecture demonstration of the issues at stake using the language of polarised light for an elementary understanding of the demonstration in 'An orthodox Understanding of the Bible with Physical Science' G E Stedman Strategic book Co. 2012 appendices. The Primary source in G E Stedman American Journal of Physics 53 pp1143--1149 (1985)

[3]   GEN A.

[4]   Courtesy of Royal Society of London

[5]   B Hunt Historical Studies in Physical and Biological sciences 16 (1) (1986) 111-134

[6]   GEN B.

2 THE SAGNAC EFFECT and its detracters

Sagnac effect

Figure 5, Georges Marc Marie Sagnac at the age of 21[1]
2013 saw the celebration of the of the centenary of the Sagnac effect in Paris.[2]

In 1913 Georges Sagnac reported a rotational induced fringe shift between counter-circulating optical beams in a small interferometer.

Sagnac also checked that no "whirling of the ether" is detected when a vertical ring is rotated. And he anticipated the use (fundamental to todays avionics) of three mutually perpendicular interferometers, to measure the roll, pitch and yaw of vehicles.

The effect of rotation was anticipated though unsuccessfully in 1893-1897 by Oliver Lodge and Joseph Larmour but not observed by them (see the Chapter 1.)

Figure 6 Sagnac's gyro[3]

We will be concerned with the case when both beams are active laser modes one laser in each direction. Both waves are amplified by passing through a suitably excited gas mixture till they oscillate. Any Rotation of the device introduces a phase shift in each beam clockwise and counter-clockwise arising from the Sagnac effect which must therefore be neutralized to retain the lasing condition. This means that the two waves spontaneously change their wave-length and so a spontaneous process reflecting the Sagnac phase shift in each beam, the two beams then acquire a frequency difference given by the formula

$$\delta f = 4\mathbf{A}.\mathbf{\Omega}/\lambda P \qquad (1)$$

where $\mathbf{A}$ is the area of the interferometer, $\mathbf{\Omega}$ the rotation rate, λ the wavelength of the wave involved, (in most of this book the wave length, will be the red He-Ne lasers transition at 633 nm). [The two exceptions to this are the consideration of the Sagnac effects for sound waves in chapter 9 and our various brief discussions of matter interferometry, we use the wavelength of the particle of equation (1). Of chapter 6.] And P is the perimeter of the interferometer the 'dot' product of the vectors $\mathbf{A}.\mathbf{\Omega}$ included the cosine of the angle between them and so will introduce the latitude of the earth-rotated interferometer, given the latitude of Christchurch (chapter3) and all these results. For the Sagnac frequency δf of a horizontal red HeNe ring laser is

$$\delta f = 317.6\mathbf{A}.\Omega/P \qquad\qquad (2)$$

One quick check on the formula (1) is that of dimensional analysis, the two lengths in the area factor are balanced by the wavelength and perimeter factors both of dimension length, and the omega factor assures that the result is a frequency as required to produce a frequency δf. When looking at the interferometer dimensions can be reduced to a single dimensionless shape dependence factor A/P^2 as illustrated by Schbois[5]. Many proofs can be presented of equation 1). The main ones are discussed in my review[6] the simplest proof which I give in that reference is to use an admittedly ether- theoretic approach to light travel in a circular geometry. That way it is easy to see phases can be matched when the light beams close on themselves after one revolution another account is at.[7] This elementary approach with some extension to non circular loops can capture the area factor which appears above . Several such proofs of the Sagnac effect are discussed in[9] coping with all such geometrical effects. Einstein synchronization in the presence of a Sagnac effect has an unavoidable discontinuity in a polygonal path say round the earth (Chapter 4).[10] General or special relativistic proofs also confirm these results[11]. One can even appeal to the Coriolis effect induced by earth rotation in the rotation frame.

One way of understanding the operation of an optical gyroscope is to recognise that counter-circulating waves set up a standing wave pattern, the beating of counter circulating waves each, governed by the vacuum speed of light, remaining as stationary as possible with respect to the local inertial frame.

Figure 7, the necklace model of a ring laser
with two counter-rotating lasing beams.

As the Earth rotates so too does the body of the ring, and so does this standing wave pattern, as and nodes of it pass by the mirrors and so the light detector, they generate a beat frequency in light emergent from the ring C-I (Chapter 7) in the audio range, giving 69 beats per second (or about 6 million antinodes per day.)

One subtlety here deserves a few words of comment. At the back of all this is the understanding that the inertial frames within which special relativity holds are themselves non-rotating. For only in them is the speed of light c. the sequence of astronomical rotations (the rotation rate of the solar system, of the galaxy and of the universe), although the size of these are known (Chapter 5), but can be ignored in discussing the Sagnac effect from more local rotations.

Near a heavy rotating mass the Lense-Thirring rotation of inertial frames (chapter 5) generates the same Sagnac effect as any other rotation in optical gyros (Chapter 7.) The inertial frames of any object may be defined not only by the absence of the acceleration of free objects but also by the absence of a Sagnac effect in a ring laser gyro placed in the that rest frame (of the object.) Another indication of the Sagnac effect is that Einstein synchronization (chapter 4) in the presence of a Sagnac effect has an unavoidable discontinuity in a path circumnavigating the earth (Peres[12]) see also another explanation of the Sagnac effect appealing to the Coriolis acceleration is particularly relevant to chapter 9. In addition there are changes in the rotational axis of the earth caused by the moon gravity acting to affect the free motion of the earth; such effects will appear in the Sagnac signal and will be demonstrated in chapter 7.

Colour shift from rotation

The effect of rotation for lasers is that the Co-rotation beams are made more red and the counter-rotating beams more blue. The argument here is that for a horizontal ring at the latitude of Christchurch New Zealand the projection of earth rotation on a horizontal ring area is anticlockwise, so during the time of transit of a clockwise light beam these figures might help as the waves must spontaneously be stretched out by the extra travel distance and realigment into a travelling wave on their recombination so satisfying the laser condition and, conversely, we will require the blue shift in wavelength, for the clockwise beam. Visualising this it may help to picture one of the following mnemonics.

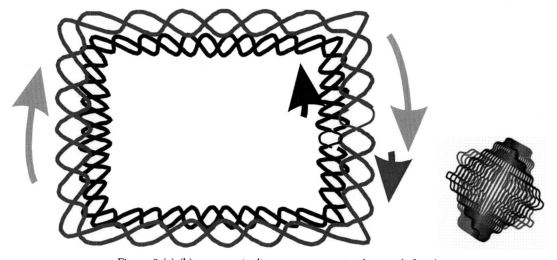

Figure 8 (a),(b) mnemonic diagrams, co-rotating beams shift red.

Tendentious discussions of the Sagnac *effect*

One enquirer Bossieman says[13]; "Is there anyone here that can explain this phenomena –the sagnac effect using Einstein's postulate that the velocity of light is constant? I haven't found any explanation to the phenomena on the Internet". My answer to this questioner, one cannot rely on the internet as I say next but I can refer you to cite one clear internet account.[14]

The reader might well be warned that the sagnac effect has spawned a literature of very dubious value. In some of this the Sagnac effect is variously painted as either contradicting or proving relativity. Most of these debates have a very long and turgid history a discussion of which is not of value here. Evidence for this on a numerous number of web sites all to easy too find on googling the words 'Sagnac effect relativity.' The Sagnac effect is variously described to be intrinsically anti-relativistic, or perhaps verifying relativity, or perhaps illustrative of some unconventional approach to the issue of the synchronisation theory as described in chapter 4. Each of these interpretations is urged with much verbosity. A recent book by Rizzi and Ruggieri[15] may serve as further evidence of such problems. Unfortunately such material serves only to inflame and confuse discussion in all related matters. Following all such threads would make this chapter would be unhelpfully tedious. The full resolutions of all such claims must remain exercises for the discerning reader with the cautions and the other material in this book.

the Tucker Review

Fortunately a good review of the book just referenced by Tucker[16] has done much to put the record straight and I would strongly recommend its careful perusal by all who might wish to see views that are deviant from orthodox wisdom clearly refuted. Tucker says for example:

"A number of such authors in the cited work believe that the results of Sagnac interferometry are in some way in conflict with the basic tenets of special relativity. Some of these concerns may stem from a paper by Ehrenfest published in 1909 arguing that special relativity was inconsistent when applied to the rotation of a 'Born rigid' cylinder. Einstein himself added to the ensuing debate that still seems to concern a number of authors in this book. It contains articles by those who argue along traditionalist lines against those with strongly held views to the contrary.

I found a number of the arguments (for and against traditional views) obscure and often ill-conceived. The essence of the case held by those with concerns about the foundations of relativity appears to be. Sagnac interferometry involves the behaviour of light observed by accelerating detectors and that it is not always possible to establish clock synchronizationwhen the clocks are accelerating.This fact then gets clouded by a lack of precision about a number of concepts that *can* and therefore should be given a precise operational meaning in any logical Discussion. Without such definitions a number of valid counter-arguments lacked impact and, for me, "the words got in the way".

"As someone who finds no inconsistency in the traditional theory of relativistic rotation, I am naturally biased against those authors in this book who work so hard to invent 'challenges' to what is manifestly one of the most elegant descriptions of nature in the physicist's vocabulary. One of the quotations in the book that I liked is perhaps most apt in this context. In a Princeton University advertisement for a particular book it states: *'a good part of science is distinguishing between useful crazy ideas and those that are just plain nutty.'* The book concludes with a series of discourses for and against the validity of the case for the anisotropy of the speed of light based on the experimentally observed Sagnac effect and other issues. Not all the book is bad, as cogent contributions by Ashby, Bel, Dieks, Mashoon, Pascual-Sanchez, Miguel, Tartaglia and Vicente 'the more reliable authors' dispel some of the mist, the protagonists remain firmly entrenched". In brief then this work propagates multiple confusions over the orthodox understanding of the one way speed of light as outlined above the validity of orthodox special relativity. One can have much discussion of supposed paradoxes in relation to the Sagnac effect. There is a useful discussion of the universality of the Sagnac equation that is why time of flight arguments for matter waves gives the Sagnac formula as reliable as they do for light waves; it is basically a consequence of the uniqueness of any synchronization scheme that is adopted.[17]' Similarly the invocation of Galilean transformations as a number of authors do can only needlessly obfuscate related matters.

One especially bad statement by the chairman in round table discussion 1 is that the Sagnac effect is a subversive tool against special relativity, the presentation here is only confused. The book concludes with a series of discourses for and against the validity of the case for the anisotropic speed of light based on the Sagnac effect (GES note all this is not the burden of chapter 4 of this book despite appearances) and other issues. One might say history has given its verdict for Einstein, not that the Sagnac effect should be seen as the experimentum-crucis to develop faith in relativity that is rather based on a portfolio of other experiments as described in any text on relativity."

Other literature comments

Related matters are unfortunately disputed in obscure corners of the Sagnac literature. I could cite a number of discussion. Part of the duties of a project leader has been to spend inordinate amounts of time discussing by email or letter with retractors of the Sagnac effect. Books like Geoff Robinson's[18] are simply best avoided. I wrestled for some time with Leo Vuky given his insistence that our large lasers showed big jumps in the Sagnac signals as the earth crossed the ecliptic in supposed conformity with his highly unconventional 'field theory'. One thing we learned the hard way more than once is never to release early data to earnest enquirers who would then put their own utterly bizarre interpretation on it and publicise the result under our name and theirs. I have been approached by at least three people who were quite sure that (despite any evidence they gave) they had a deep understanding of laser gyroscopes and a vital new insight of huge commercial value. But they could never give me any hint as to what this insight was until we had signed a nondisclosure agreement etc. Nothing ever happened. I guess they soon realised I did not hold the purse strings of the industry or of their imaginations.

Malkin[19] makes the almost incredible (for a follower of relativity) statement that the group velocity transforms under Lorentz transformations as a particle velocity. A proper account can be found in[20]. As Tucker noted the invocation of Galilean transformations as a number of authors do can only needlessly obfuscate related matters.

While Sagnac and Lodges work was concerned with studying phase shifts in optical interferometry (chapter 1), Michelson once estimated the likely size of an interferometer to detect the rotation of the Earth around the Sun as, 100 square kilometers Michelson once used an interferometer with an area of order one square kilometer to measure Earth rotation He performed this experiment with an actual area (0.21 km^2) in Clearing, Illinois in 1925 (see Chapter 7).

Macek and Davies' historic experiment on a 1-meter square rotatable ring laser

This Helium-Neon ring laser built in 1962, has set the technology for recent advances and has made the Sagnac 'frequency splitting' effect well recognized. The phase shift of the old optical interferometers has become automatically transmuted by the lasing action in each beam into a frequency shift between the counter-rotating beams, and since frequencies are measureable to high accuracy this enables very precise measurements of rotational effects.

The light beams from the counter-propagating waves as they leak through any mirror even through the best at pico-watt level and on the superposition of the beams they beat. The interference fringes formed move at constant speed past any light detector. This converts the Sagnac observable into a frequency in an active (lasing) device which has had enormous repercussions in the optical gyroscope industry- this is built into equations 1,2 of this chapter. Even though the speed of light is proverbially high, severely limiting the angle through which the system can revolve during the transit time, frequencies are accurately measurable down to very low values, say a fraction of the optical frequency such as the He-Ne laser frequency (474 THz). this is a big advance over an old optical interferometer. Laser gyroscopes are currently among the most sensitive gyroscopes available. cleanliness and ruggedness and the accuracy of electronic frequency counting in a purely optical experiment, the use of no moving parts, all help to make the mean time between failures greatly extended over mechanical gyros. The body language of the gyro industry was in-compatible with the larger ring lasers of chapter 7. This advance was widely thought impossible because of Quantum noise, for this sort of reason other better equipped groups had not attempted our kind of large-ring laser program and we had the field to ourselves. The serendipitous advance in mirror coating technology in the '70's and 80's that permitted optical cavities of unprecedented high quality factor to be generated. This was also largely unreported in the scientific literature, the techniques being highly protected commercial secrets. The beams both travel at a fixed speed c which nowadays is defined to be 299792458 m/s (that is the standards of length and time are defined), for that immediately derives from the Einstein postulates for special relativity. Nowadays, we think of detecting phase shifts $\sim 10^{-40}$ radians from the Lense-Thirring effect –(see relativity chapter 5) due for checking in the ``Schiff gyro" experiment - [21] the precision required may be attainable with ring lasers, chapter 7.

Mirrors

Mirror quality may be specified by the power reflectance R. One key result of the commercial and military development has been a vast improvement in mirror quality, ``five-9s quality" (R= 99.999%) and now even ``six-9s" (that is a 1 ppm total loss of power in reflection) characterizing the ``super-mirrors" now available.

The fact that C-I worked as well as it did was sufficient to prove the principle of moving to large areas than the typically 60 cm^2 navigation lasers was feasible. Our experiment was made possible by the donation of the state of the art mirrors with typically 22 layers of TiO_2/SiO_2 (given to us by a kindly mirror manufacturer to whom Hans Bilger appealed for pure research support[22].)

Figure 9, a super-mirror in natural light

Of course this photograph is taken in ambient light and cannot display their reflection capabilities at the design wavelength namely for the HeNe laser line at 633 nm. the $\lambda/4$ thickness layers are chosen to match for He-Ne use for which these mirrors can have up to 99.9985% reflectance. As mentioned the huge secret technological developments in the '70s and '80s made ring lasers ever more sensitive to rotation. Today's super-mirror reflects 99.9999% of the light it receives. Much of my early correspondence with Professor Bilger at the University of Stillwater OK, USA and discussions the details of the theory of multilayer dielectric coatings using alternating layers of TiO_2 and SiO_2 of and the implications of these for Zerodur mirror substrates were carefully estimated using the refractive indices . These supported mirrors with about 25 dielectric layers with for total losses of the Order of 10 ppm; and was a surface roughness of 0.1 nm RMS according to Tony Louderback with whom Hans was also vigorously determining specification and quotation. Eventually appropriate mirrors arrived late 1987 (we paid for the mirror blanks) although we could not get all data on his mirrors . We always found manufacturers loath to quote key parameters. This led to much discussion involving the outstanding optical technician of the New Zealand DSIR although Invar was considered and indeed tried for mirror holders but in the initial version stainless-steel was used. Equally there was much discussion over the material for mirror blanks Zerodur or SiO_2. This was partly in the hands of the mirror manufacturers C-I and all subsequent lasers actually used Zerodur for the mirror blanks as it takes the necessary high polish including and Ion beam smoothing. This is despite the obvious optical loss of the material (indicated by its orange colour of Zerodur) and the radio frequency loss which might be more serious given the proximity of the mirror boxes to the gain tube and r.f. or radio frequency coil that excites the He-Ne gas.

Such measurements indicate why ring lasers are now standard inertial navigation sensors in military and commercial aviation.

In pacifist style, the Canterbury ring laser program ploughs these advances back into pure science. We collaborated strongly with the TUM Technical University of Munich and the German Federal Institute BKG Bundesamt für Cartography und Geodäsie (Professor Ulrich Schreiber being the leader of their ring laser programmes.) In improving the quality factor of the cavity, the operation with regard to lock-in, frequency resolution, signal/noise, backscatter-induced pulling of the frequency etc. is improved at least in proportion.

The history of laser gyroscopes is a curious century-long thread of ingenious and outrageous yet unabashed speculation, brilliant innovation, fierce commercial competition including billion-dollar lawsuits, maximal military secrecy and cold-war rivalry. The last of these in particular were the moving force for many dramatic but little-known scientific advances in the 1970s and 1980s in the drive for better precision and commercial and military deployment. Some (but not all) of these advances are now declassified.

When we first started, to get super-mirrors you had to have friends; money wasn't good enough. (It is different now; now you need money.) Our original manufacturer Louderback was most generous to us, Louderbacks mirrors had losses of 14 ppm. His methods were state of the art and possible featured in Litton-Honeywell suits and countersuits, and Louderback's firm no longer operates. In those better days (1989) we once got to the stage of filing a joint patent application within New Zealand relating to the manufacturer of mirrors with minimal birefringence We became interested in birefringence effects in consequence of realizing that the polarization of the laser was strongly affected by any twisting of the ring out of an exact plane. Typically the polarization changes are dictated by the difference in reflectivity in different polarizations, about 150 ppm, and the change is from linear to circular polarization over 150 ppm of a radian (the standard unit of angle and about 57 degrees). We thought that this had the potential to make a tilt-meter that could measure a few hundreds of pico-radians - the angle made by a needle at the distance of the UK from NZ. The fact that reagents existed which could strip off atomic layers in a controlled manner made such effects controllable, so that the birefringence could be reduced to a 'quantum limit'. I was urged by others in the group to patent this idea and struggled very hard, for a year before the provisional patent expired, to interest Litton into buying this patent without success. Our idea has not been used yet as far as I know.

Such mirrors involve say 22 pairs of quarter-wavelength layers of materials such as SiO_2 and either TiO_2 or Ta_2O_5 with differing refractive indices. These are sequentially evaporated onto a well-prepared substrate by ion deposition. A surface half-wavelength coating helps protect the super-mirror, which is then surprisingly robust. It is critically important that the substrate be ultra-smooth, since any 'hill' will make its presence felt throughout the superimposed layers and so limit the mirror finesse regardless of the reflectivity of a small section. With controlled ion bombardment by argon and nitrogen for example, substrate surfaces can be prepared with an r.m.s roughness well below the molecular diameter; one super-mirror in our possession was measured to have an 'rms surface roughness' of 0.016 Angstrom. Obviously, the ion deposition process is also quite critically important.

An example of the commercial pressures is given in the award by a USA jury in 1993 of US$1.2 billion in damages in a lawsuit, Litton Systems v. Honeywell Inc., in connection with the ion-beam mirror coating technology. In March 1996 this jury verdict was reversed by a judge because, the patent being declared invalid, the license agreement became unenforceable "there is a total absence of evidence that the mirror manufacture ever actually used any Litton trade secret". Present commercial methods could still be improved; we are still several orders of magnitude from the limits set by the intrinsic absorption of the mirror coating materials. It is therefore reasonable to speculate about "parts per billion" and not merely "parts per million" super-mirrors[23].

Particle interfero-metric gyros

Referring to equation (1) of chapter 6 The bigger the mass and the slower the particle, the larger is the associated de-Broglie wavelength and so from equation (1) of this chapter. The more sensitive is the resulting Sagnac gyroscope.

Electronic gyros

One can build gyros from two Josephson junctions between superconductors for example gyroscopes have been made by building interferometers from these components. [24] Other examples are mentioned below see also Franz Hasselbach[25].

Neutrons are heavier than electrons, and with care (and obviously expense!) a neutron beam from an atomic reactor can be slowed down into a thin coherent beam of waves. They can then be made to interfere by diffraction off parts of a perfect silicon crystal on account of their deBroglie wavelength (Chapter 6). All kinds of quantum tricks have been demonstrated in these devices. Earth rotation was measured in a neutron interferometer of size a few square centimeters in 1989[26].

Super-fluids helium atoms at cold temperatures can form Josepheson junctions. And gyroscopes have been built from these[27].

Atoms are heavier than neutrons, and within the last few years it has been found how to cool them down (and slow them down) to a billionth of a degree above absolute zero[28].

The Sagnac phase shift for matter waves has been verified with accuracy on the order of 1% for neutrons and electrons (Hasselbach and Nicklaus[29]), and to about 10% for atom interferometers.

GPS has been an enormous help to us at Cashmere. With an aerial at the door it is used to stabilize the frequencies of generation of radio frequency generators and the frequency counting circuits for detecting the Sagnac fringes for all the lasers, and all such comparisons.

I mention some of the gambles taken in building our larger lasers:

Single mode

One serious requirement was if at all possible to achieve single mode excitation of lasers: That is to having say 7000000 and not 7000001 wavelengths around the perimeter, otherwise the different Sagnac frequencies for such modes would beat in the detector. We met this issue successfully in C-I, C-II and G0 while in U-G1, UG2 projects we gambled the project on success. Our basic strategy was starvation of the unwanted modes: we would lower the power of the plasma excitation till the laser operated on only just one such longitudinal mode. This strategy worked far better than we had good reason to expect. In a ring as big as C-I or C-II, there are more than 10 modes in the gain curve. In G0, whose perimeter is 14 m, the FSR is 21 MHz, and there are 50 modes that could lase! And there are possibly transverse modes as well. Nevertheless the starvation technique works, thanks to mode competition - the likely winner stands higher in the pecking order, and tends to grab the energy in the other modes. Probably this was the biggest single success of C-I, and of G0, justifying the concept and relieving a major worry of the much larger investment in G. Another trick helping this which we owed to Bob Dunn was the use of higher pressures of the He-Ne gas than would normally be recommended, the Doppler or pressure broadening so induced in the laser line changed the dynamics of the mode choice within the gain curve and allowed an extra control. If other modes were excited that could be useful, the beat of modes with modes gave a signal in the radio frequency regime allowing precise measurement of the ring perimeter.

Locking

When two oscillators are even slightly in contact, their frequencies become affected; they try to lock together. Christian Huygens saw this in the 1629-95. Sick in bed, he watched two pendulum clocks on the wall -- they ticked together, locked in frequency. As one pendulum swung, it moved the wall slightly and so moved the other pendulum through the wall movement. Whether two oscillators lock or not depends on a judicious balancing act between the amount of coupling (in our case the proportion of backscattered light that gets into the other beam) and the mismatch of the natural frequencies (in our case the nominal Sagnac frequency splitting). Ring lasers the size of an aircraft gyro lock under Earth rotation; the Sagnac frequency mismatch is too small to shake the two laser modes free. A complicated and highly classified system of dither -- of shaking the gyro, so as to shake the modes free -- is vital. The second gamble we took with C-I was that Earth rotation alone would be fast enough, and the mirrors good enough, for C-I to unlock without any special tricks. It worked if the mirrors were good enough.

Even if the system does not lock, the actual frequency difference is less than the nominal Sagnac value: the frequency is pulled. This is a basic reason for the limits on the accuracy of the ring laser gyro. The "Earth rotation" signal actually varied by 10%-20% in C-I, depending on the dirt on the mirrors. Even in C-II, in its initial form, it varied by up to a few per cent in the worst case.

It was found very roughly in C-I and quite accurately in C-II that frequency pulling effects in such big rings were much worse than had been fully expected. In making C-II, the argument had been: if the mechanics is stable to parts in 10 million, the Sagnac frequency --which depends on things like area and perimeter -- should also be stable to parts in 10 million. It wasn't - it drifted by parts in 10000 or more.

Solving this problem has been a major challenge for 1997-1999 and beyond, and in retrospect it provided a full justification for the building of C-II as a test for G.

Frequency pulling depends on the details of wave addition and cancellation. It was recognised to be caused by backscatter and heavy influence by mirror imperfection and cleanliness. Dust spots in two mirrors can each backscatter one beam into the other, and it turns out that the degree of pulling is very sensitive to whether these two backscattered beams add in phase or out of phase. And that itself changes dramatically with pressure on the block as a whole. The dynamics of his phase change on laser performance requires an analysis of the lasing mode equations, as laid down by other authors. My approach has been traditional: indeed. I have in seminars used a Perspex model to understand the wave interactions and the interconnections of the phasor vectors describing the effects on the time evolution of the lasing modes.[30]

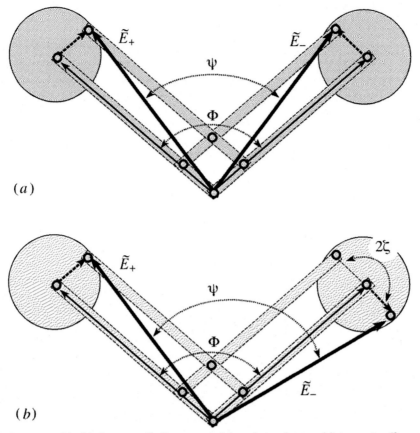

Figure 10 (a), (b) diagram of a Perspex model to show phasor addition as it affects
frequency pulling from backscatter.

Phasor diagrams for describing pulling, based on backscatter. One may regard the broken lines as a mechanical framework with the indicated pivots. Indeed the above diagram shows is a model of a Perspex display for an overhead projector. A proportion (dashed vector) of one beams phasor (thin solid vector) is added to the other, indicating the time development of the final phasors $E_+ E_-$ (thick solid vectors). In the absence of pulling, Φ is the phase difference between the two beams increase in time linearly; . $\mathcal{Y}$ is this phase adjusted for pulling, and itself reacts on the dynamics of. (a) Dissipative coupling (b) Conservative coupling these cases distinguishing different time reversal properties of the scatteres.

For details see [31] the full story on the effects of backscatter is still being developed and I leave this discussion here.

We found by hard experience that the best way forward was not by stabilising the perimeter and the optical frequency, in Ulli Schreiber's commissioning year 1997. For C-II this was done explicitly by mixing the C-II beam with an iodine stabilised laser detecting using a high quality Fabry Perot and applying a correction signal to the C-II Zerodur block to change the mirror separation, at this stage the effects of ambient pressure changes with the weather mainly on the more flexible mirror mountings could be uncontrolled.

Although expanding the ring changed the mirror separation, the wavelength also changed and in the same ratio, so that the phase does not change). C-II was eventually enclosed in a pressure vessel to reduce the effects of backscatter. This made possible a 100-fold reduction in Sagnac frequency drift in December 1998.

It is described in chapter 7 under UG1 how a major discovery was made with our larger ring lasers in that they show up the daily effects of the moon on the rotational axis of the earth.

We got very interested in the control of mirror birefringence following the understanding that reagents can be used by the mirror manufacturer to remove one atomic layer at a time from the surface of a mirror film. In principle such methods can be used to give very sensitive tests of the planarity of the ring by monitoring the mode polarisation.[32]

The behaviour of polarisation of a beam of the light in a ring of mirrors can be understood to some extent from the idea of parallel transport. Hold your arms out parallel and in front of you (their line representing the direction of the beam), with your thumbs upwards (representing the polarisation), then open your arms horizontally till they are pointing in opposite directions out from each side of your body (the thumbs are still parallel), bring them up above your head so that the thumbs are opposite, then and down to the first position away from your chest– again the thumbs are naturally opposite. So the natural (locally parallel) transport of a polarisation vector round a ring in three dimensions can change the relative polarisation differently depending on the path that is taken. In fact the overall angle of rotation of each thumb is twice the solid angle traced out by your arm as you trace the above circuit. This is an example of the Berry phase, whose full significance in quantum theory has been appreciated only since 1982. Michael Berry showed that when a particle goes slowly on a full circle round geometrically special points, it can acquire a phase shift which is dictated by the geometry of this loop. Light beams that once were in phase when sent on a twisted path can twist their polarisations so as to interfere [33].

A net polarisation rotation for a beam travelling one circuit of a buckled ring means that the ring can no longer lase in pure linear polarisation; So its choice of polarisation is elliptical, and measurement of this ellipse turns out to be a very sensitive measure of the buckling[34]. Lewis Ryder investigated if the Sagnac effect depends on polarisation we might be able to detect torsion in space I hope this little discussion illustrates how many things there are in science to be understood on careful investigation. Giving the scientist a few mirrors and we can find all manner of mysteries.

.It is important to appreciate just how staggeringly large or small some of the numbers are for optical beams in ring lasers, the basic He-Ne laser frequency can conveniently be blown up by 12 factors of 50, over which range the associated wavelengths from the visible to astronomically interesting sizes.

In the very earliest days of this project Hans Bilger had warned me that people would not believe the extraordinarily small scale of the effects ring lasers could explore. He later reminded me of that as recorded in Lydia's book on Brian Wybourne[35] one of the reviewers of our first papers 'flatly' corrected our high resolution measurements imagining that our mHz (milli-Hertz) were really MHz (mega-Hertz) thus introducing an error of a factor 10^9. So the next two posters were drafted in an effort to convey this huge scale in the potential observables.

Orders of magnitude

In the first poster below the frequency scale is stepped up by several factors of 50 as one reads from left to right, the wavelengths change accordingly as depicted in the central bar. In the second similarly we blow up the earth rotation Sagnac effect by factors of 10 up to 11 times and consider the sensitivity of such a system to the corresponding Sagnac effects.

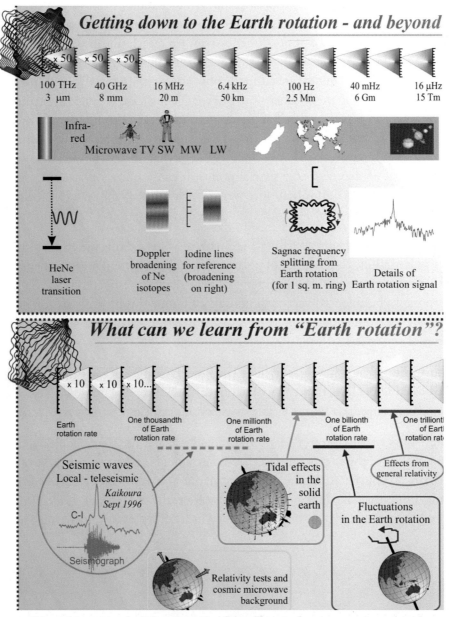

Figure 11, posters depicting the scale of the effects in frequency and wavelength
and so physical terms accessible to large ring lasers.

The range of effects that may be studied is over dimensions up to planetary size. The number of physical tools with such precision is limited not: even the precision of the Mossbauer effect is comparable with ring lasers for frequency discrimination. Of existing devices G is certainly detecting fluctuations in the earth's rotation rate and so the potential advertised in this old poster for detecting effects at the level of one billionth of the earth rotation signal.

When C-II was commissioned it was estimated that the coherence length of the 633nm photons corresponding to the He-Ne Laser line centre on the quantum noise limit was greater than one light year.

Laser Stability

One curious little story may interest you, a mirror mystery about the planning of our unprecedentedly large lasers might be slipped in here. It is obvious on a little thought that the optical stability of a laser setup requires non trivial placing not merely of the mirror positions and choice of their radii of curvature. Some miracle unexplained to this point must be going on if a beam of light once sent on its way round a polygonal ring (or even a linear ring) yet maintain its required ray path or axis without significant deviation after a few reflections so that the beam walks off the mirrors. The mirror positioning can never be exactly perfect, why is it that the inevitable amount of imperfection that would kill lazing action? This problem is particularly obvious for a ring of flat mirrors. The problem of making a ring laser must seem nigh insoluble, the reason for the curvature of the mirrors is to offset this problem steering the beam in a more sustainable direction partly though the geometry of reflection but more particularly through wave diffraction effects and ensure that the 'Gaussian' laser waves of light are redirected into is not near perfect directions so that deviations average out to unimportant and tolerable amounts. This problem is fully worked out, as it would seem to be in many texts and papers on lasers, for example[36] (one which had itself mean appropriated by laser physicists from parallel problems in microwave optics). Hans Bilger in designing C-I had presented this to me as if it were a new problem needing solution for such a laser as C-I. True Hans did at times muttered to me about 'ABCD' matrices but I was ignorant enough to ignore it. This gave one well known stability condition. By accident more than by design I fell on some novel aspects of this problem. One accident was that as a solid state physicist by training I was ignorant of the work that had been done over many years on laser stability. So I naturally tackled the problem in a totally unconventional manner[37] simply by ray-tracing from the basic equations for reflection and I came up with a new result, a second stability condition that had to be satisfied over and above the conventional one of laser lore. In time I learned the official way of discussing laser stability using ABCD matrices (the best way through having to teach it) but then found that I had a very serious mathematical job to reconcile the different approaches. The standard approach- to stability, and the one I had developed, actually gave significantly different results complementing each other, Some of my more conventionally trained optics colleagues took some convincing of the issues at stake. It turned out that the reconciliations needed theorems about determinants in mathematics that were unknown to me, and a kind mathematician helped me out here. This is no place for mathematical details. But the situation lends itself to a verbal description of the issues which I would like to offer the reader. Full details which were particularly important for planning our large lasers can be found in[38] a paper in which my earlier ignorance of the ABCD matrix method was revenged by offering a new matrix method which I called ABCDEF to cope with other displacements DF of the mirrors through misalignment, also treated to first order, as were the effects of mirror curvature and separation in the standard method.

Thus two stability conditions arise then which I like to think of in anthropomorphic terms. The standard test is like guarding against the possibility that a well-raised child will kick off the traces and leave home –that is, just as a laser beam will eventually work itself out of the laser cavity after multiple reflections because of the cumulative effects of mirror misalignments.

I think the other test is more like the condition that a child raised in a less than perfect environment (i.e. with significant displacements of the mirror poles from ideal positions) will nevertheless make a go of its environment and make a fist of life. When both tests are available one can get a fully informed set of conditions for the stability of a laser cavity design.

Endnotes

[1]   M. Quintin, »Qui a découvert la fluorescence X?«,
Journal de Physique IV, Colloque C4, Supplément
au Journal de Physique III; **6**, (juillet 1996); pp. 599 - 609

[2]   10/10/13 at the Fondation Simone et Cino del Duca of the Institut de France.
"Comptes Rendus de l'Académie des Sciences vol **157** pp. 708-10".

[3]   Wikipeakea, article sagnac.

[4]   L Ryder Introduction to General Relativity (Cambridge 2009)

[5]   Schbois E O 1987 IEEE J Quantum Electronics **QE-2** 299-305

[6]   GEN, A

[7]   http://mathpages.com/rr/s2-07/2-07.html

[8]   GEN D

[9]   GEN A

[10] GEN D

[11] one recently published general relativistic proof is that of L Ryder Introduction to General Relativity (Cambridge 2009)

[12] Peres 1978 Physical Review **18** 2173-4

[13] http://www.physicsforums.com/showthread.php?t=3271

[14] http://mathpages.com/rr/s2-07/2-07.html

[15] GEN E.

[16] Tucker General Relativity and Gravitation **37**, 1159-1161 (2005)

[17] See for example G. Rizzi and M Ruggiero General Relativity and Gravitation **35** 0001-7701

[18] from book Relatively Simple Relativity myth and mystery dispelled Geoff Robinson 2nd Edition 2008 Published by Lulu.com

[19] M Malkin Phys Uspecki **431229**-01552 (2000)

[20] G E Stedman Amer J Physics **60** 1118-1122 (1992)

[21] *Physical Review Letters* (20011) **106** (22): 221101.

[22] Tony Louderback of Ojai Research.it is a great relief that we can finally and gladly record our very great debt to him. In post C-I years and so for the larger lasers discussed below rest-restrictions have eased super-mirrors have became commercially available.

[23] H R Bilger, P V Wells and G E Stedman, *Origin of fundamental limits for losses on reflection at multilayer dielectric mirrors,* Appl. Opt. **33**, 7390-7396 (1994).

[24] GEN F.

[25] Scanning Microscopy Vol. **11**, 1997 (Pages 345-366)

[26] Neutron Interferometry Lessons in experimental Quantum mechanics H. Rauch and S Werner (Clarendon Press Oxford 2000)

[27] http://www.physics.berkeley.edu/research/packard/current_research/nielsweb/phase_slip_gyro.htm

[28] http://calyptus.caltech.edu/qis2009/documents/kasevichQIS0409.pdf

[29] F.Hasselbach and MNicklaus Phys. Rev A**48**143-51 (1993)

[30] GEN A

[31] GEN A

[32] H R Bilger, et al., Geometrical dependence of polarisation in near-planar ring lasers Opt. Commun. 80 133—137 (1990).

[33] http://en.wikipedia.org/wiki/Berry_connection_and_curvature

[34] H R Bilger, G E Stedman and P.V. Wells, *Geometrical dependence of polarisation in near-planar ring lasers* Opt. Commun. **80** 133--137 (1990).

[35] Lydia Smentic: Brian Garner Wybourne Memories and Memoirs adam marszalek Totun (2005)

[36] H. Kogelnick and T. Li applied Optics **5** 1550 (1966)
Rigrod J. Appl. Phys. **34**, 2602 (1963)]

[37] H Bilger and G E Stedman Applied Optics **26** 3710 (1987)

[38] Currie Stedman and Dunn, Applied Optics **41** 1689 (2002)

3 The cashmere cavern

Longitude 172.672 E   latitude 43.57694 S

In a pacifist style, the Canterbury ring laser program ploughs these advances in gyroscope mirror technology (Chapter 2) developed in a strongly military environment back into pure science this was done at a military installation the Cashmere Cavern. While reconciliation with former enemies is evidenced by a collaboration strongly with the Technical University of Munich TUM and the BKG German Federal Institute for Cartography and Geodäesy (Professor Ulrich Schreiber being the leader of their ring laser programmes). Our laboratory is 100' underground at Cashmere, Christchurch, New Zealand. (This facility was built for a World War II as a command bunker) in the event of Japanese invasion. Its mechanical foundations and temperature both very stable.

Figure 12, location of Cashmere
Cavern in New Zealand

The location of the ring laser laboratory at Canterbury University was a major headache. Initially it was set up on the (top) 8th floor of the Physics Department building which like all campus buildings essentially floated on a concrete pier in the shingle of the alluvial Canterbury Plains where it successfully monitored not only earth rotation but also was sufficient to confirm the principle of a device of this type, the motion of the building in the prevailing föhn winds from the north-west, on the west facing building, also the differential effects of solar heating on the east and west facing faces as the day progressed, lift motion, and even people using the stairs in a central service block which serviced Chemistry and Physics Departments. Clearly something freer of vibrations was needed and it would be off campus.

Alternative sites investigated

Initially we looked at some military tunnels overlooking the container wharf Cash in Quay in Lyttelton the port of Christchurch. We also looked at an abandoned seismological station at Gebbies Pass at the head of Lyttelton harbour was a deep concrete pillar was an attractive mount for a table-top experiment . However as room was limited and being under the shadow of the tall aerials of some local radio transmitting stations the environment here was not ideal electrically. And the location was comparatively remote from the city and university.

Then quite serendipitously a new option presented itself. The Cracroft estate was originally founded by Sir John Cracroft-Wilson in 1854. Sir John was a holder of the Star of India and both he and his descendants have been prominent figures in Christchurch society. Gordon Ogilvie's masterly 'Port Hills of Christchurch'[1] gives an account of this outstanding pioneer.

Cracroft built a Mansion of a home on Cashmere Hill, Ogilvie has good photographs of the magnificent old mansion, designed by renowned architect Samuel Hurst Seager. It was Cracroft who brought rhododendron seeds from India for the Ilam gardens, which are now under our university management. In 1942 his homestead and adjoining land was commandeered as the Southern Group Headquarters, with the homestead for officers accommodation, and works commenced on a large cavern. Located just over 100 ft below ground, the three main caverns of the bunkers together form a U shape. The largest of the caverns is a massive 7 m high, 10 m wide and about 40 m long. The main entrance to the caverns for officers was via a set of stairs that descended 100 ft from the cellar of the Cracroft-Wilson homestead. Staffs of other ranks entered the bunkers through access adits in the side of the hill.

The caverns of the bunkers were lined with reinforced concrete. Flt Officer Andrew Gray was stationed at Southern Group headquarters for a time and remembers the prefabricated concrete linings arriving slowly up the hill from Mt Somers on Burnett's Transport trucks and being lifted into place from the adit railway by hydraulic jacks.

The proposed layout for the bunkers was to be similar to Winston Churchill's war cabinet rooms in London. For example the largest and most well roofed area at its south end was envisioned as the general mapping and control room for conduct of the war. where progress could be monitored by all staff it was therefore in a particularly secure location under a heavy concrete covering. a mezzanine floor would aid the inclusion of a radar room a 'cypher room' and a 'combined ops room' and with kitchens, toilets, and ventilation and communication systems. All this suggests that Southern Group were preparing to be equipped for a long stay underground.

By April 1943 the bunkers were nearing completion but work came to a sudden halt due to changing fortunes in the war. The Japanese invasion seemed unlikely, meaning that Southern Group and the bunkers at Cashmere were no longer necessary. The Public Works Department had been working on the bunkers 1 ½ years at a total cost of £35,204.

Figure 13 interior view of the main cavern.

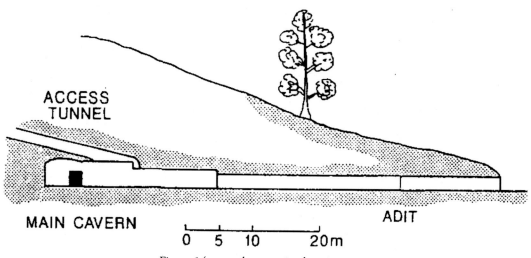

Figure 14, tunnel system in elevation

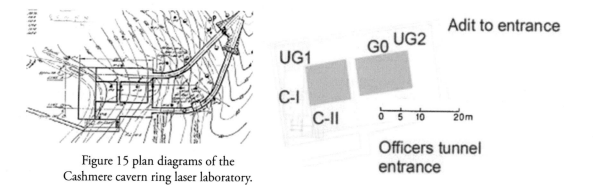

Figure 15 plan diagrams of the
Cashmere cavern ring laser laboratory.

The military cavern at Cashmere, is in a southern and hilly suburb of Christchurch city. It's existence had been a military secret but was publicly revealed 1988 though a chain of events, which I briefly review[2]. Countries like Australia New Zealand was clearly vulnerable to Japan expansionism in the Pacific theatre; Japan had opened the Pacific war by bombing the USA Navy at Pearl Harbour Hawaii on 7 December 1941. Hitler disastrously announced he was at war with the USA and the USA came into World War II. The naval forces of USA and Japan clashed decisively in the Coral Sea. Military action grew close to Australia including the bombing of Darwin and then Japanese submarine action in Sydney Harbour. The coastline of New Zealand was clearly under Japanese observation and the threat of Japanese invasion could not be discounted. A considerable number of military installations were hurriedly built, several flanking the hill and ridges overlooking strategically significant places like Lyttelton Harbour near Christchurch and indeed other military installation existed further afield. It turned out that in each major centre of New Zealand a major installation was built to retard any Japanese invasion. It was as a part of this defensive effort the Cashmere cavern was drilled and blasted by miners out of solid volcanic basalt rock about 100 ft underground, 42 meters' long 21 m wide and up to 5 m high. A ventilation shaft 60' high dropped from a lawn of the Cracroft homestead descends to the officers tunnel linking the basement of the old house to the officers entry, (we were told that coal miners from Colgate had been appointed for that side tunnel excavation).

Although the Cracroft Caverns were never fully completed- never fully roofed or lined for example, their story does not end there. By December 1944 the Army and the Navy had left Southern Group and the Cracroft-Wilson estate. An obscure story tells that on the end of the war the officers had a party the weekend before the army was to be returned the homestead to the owners and a fire destroying the homestead that night, the so its architectural gems were lost although the family heirlooms had been secured.[3] I suppose Cromwell's troops did no better when billeted at Hollyrood Palace in Edinburgh. The Cracroft-Wilson house was replaced with a ferro-concrete structure and is still there today as a private dwelling. It is no longer in the family.

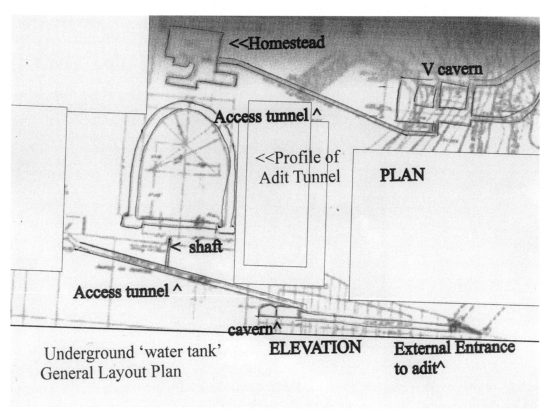

Figure 16, example of military plans

This bunker was intended to serve as a potential command post for Army Navy and Air Force command staff, if only for the early days of a real Japanese invasion. On the surface there was nothing spectacular about Southern Group headquarters at Cracroft, they looked like a typical military base with a Mess Hall, Airman's Quarters, etc. Underneath the ground however there was something out of the ordinary, a top secret set of caverns and tunnels designed to be the underground nerve centre for the military defence of the South Island. The job of blasting the caverns into the hillside started in 1942. The work was undertaken by the Public Works Department under instructions from the military. The whole operation was conducted under the strictest of secrecy and constant vigilance of the military sentries. The secrecy was sufficiently good that although nearby residents were aware of the blasting they never knew the intention of the project. Even world famous mystery writer and a local resident, Dame Ngaio Marsh. mentions it in her autobiography[4]"We could hear and feel blasting under our hills where, it was rumoured munitions were being secreted."Roofing was installed and spoil removed using a light railway looping in to the cavern through one of the entrance adits, and out of the other.

Figure 17, west adit entrance

While the bulk of the cavern is excavated from volcanic basalt these adits are largely driven though the loess deposits which coat the port hills, i.e. windblown glacial dust driven over the Canterbury plains by the föhn north wester winds. These adit walls were coated with sprayed concrete 'shotcrete'). A loop rail linked these West and East adits to their entrances one for egress at the other for egress and removal of excavated material. some rock were trucked away. Some used to form the present access road some used to fill portions of the nearby valley and even created cliffs of up to 30 ft.

Also the railway was also used for bringing in precast concrete cavern lining pieces these were trucked in from Mt Somers and were possibly the first examples of pre-stressed concrete construction in New Zealand. Railed into the cavern and lifted on hydraulic jacks to roof height. In building our ring lasers we found that the accuracy of the military engineering was outstanding, for example in making the walls vertical to 1mm accuracy when it came to using their vertical walls supporting these roof beams this was a major asset to us. It would have been easier for us if the military work had not stopped when it did, for this part of the job was never completed but as the Japanese threat receded, on 7 April 1943 work on the cavern was called off[5].

Figure 18 Bernie Bicknell

We have met one man who was assigned for radar operations in the cavern.

Some military use was made of the caverns; for example, a local resident, Bernie Bicknell, was a radio operator in the caverns during the war. One of our many more knowledgeable vocal visitors, Bernie had many vivid memories from war days in UK one of the "Chain-Home" coastal radar operators in the UK defences near Dover when WAF officers were attacked by Stukas. He could recall early days of the cavern building when he visited us. Latterly he built a large pipe organ in his cashmere home.

Jeff Field

The military prevented public knowledge of the project, regarding it as a military secret into the late 1980's when the story of the cavern first broke publicly. In 1988 an inquisitive television reporter Jeff Field learned of the existence of the tunnel complex.

Figure 19 Jeff Field (at the foot of the officers tunnel)

Jeff is now Registrar of the University of Canterbury. His mother had been a ward clerk at the adjoining Princes Margaret Hospital and at her retirement function. Jeff, as a good reporter, quizzed a gardener about any possible items of interest in the area. As a consequence he learned of the caverns and visited the Ministry of Defence, he gained access to their library, where he found full plans of the structure. He also made the decisive discovery in that library of a photograph from the Christchurch Press newspaper (the major publicity organ for Christchurch) of the machine used to help excavate the main parts of the cavern. When the Ministry of Defence later told him in 1988 that he could not publicise the material because the existence of the cavern was still a military secret his response was to point to this photograph—the secret had already been blown in the local media. The secret of such a big project had actually been kept remarkably well.

So with the agreement of the Cracroft family Jeff, who by now was Chief Reporter at Television New Zealand, arranged to cracked open the concrete top to the ventilation shaft. With the help of an abseiling group Jeff's colleague Bill Cockram descended the 60' ventilation shaft into the centre of the cavern facility. News items featured on the national television news and on the regional programme "The Mainland Touch". One of the University of Canterbury technicians, Morrie Poulton who saw the broadcast said to himself that would be a good place for our ring lasers, a secluded location mechanically and electrically quiet. As the Vice Chancellor later told him it pays to watch TV. Morrie immediately started negotiations with Bill Wilson of the Cracroft family for the university to be allowed access to the cave for a laboratory.

A side lawn to the present homestead then gave access to the 60' ventilation shaft through which we and TVNZ, first gained access to the cavern. Its floor is about 100' below the lawn (in which the ventilation shaft was sunk) the adit entrances had been sealed over after the war by owner Bill Wilson,

He put a concrete lid over the ventilation shaft. The officers tunnel was sealed off by a brick wall not far from the house or from the foot of the ventilation shaft. At one stage we had severe water leak problems which led to an investigation of this tunnel above the brick wall this required opening the tunnel at the old cellar of the homestead.

Figures 20 (a) above original officers tunnel near house cellar (b) near foot of ventilation shaft

Officers tunnel near bottom of ventilation shaft. In recent years (2004 on) there has been a significant problem with excessive water run off from the hill above or possibly pipe leaks resulting in water finding its way down the officer's tunnel into the cavern.

The Wilsons had planned to reseal the shaft and after its opening had slung a few railway sleepers over the dangerous hole. This proved insufficient to deter the boys in the area from moving them around over the following weekend. Apprised of the problem Morrie rapidly arranged to pour a new concrete top with a lockable opening hatch. That certainly fixed the vandalising problem for the moment (we had plenty of other vandal problems later if less dangerous when we opened the adit below). In those early days this was our only access and we abseiled down fortunately a feasible test of strength of several of the technical staff who had conducted pilot temperature and humidity tests. I was lowered down once and pulled up again by these abseiling-savy technicians along with the then head of department Prof Brian Wybourne.

Figure 21, Brian Wybourne on abseiling ropes

Brian Wybourne above the 60' drop. He was lowered then later pulled up as I was. The technicians involved jokingly discussed when they should ask Brian as Head Of Department for a pay raise. Hanging as he was at their mercy.

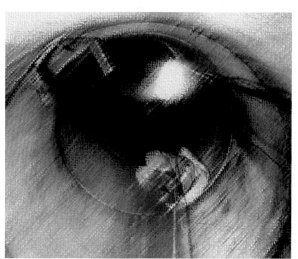

Figure 22 Clive Rowe descending by more conventional abseiling techniques.

mentions of the cavern in a geologic context are [6] [7]

An eloquent chapter on our use of the cavern is in Mark P Silverman's book p. 184 "As I stood in an unlighted passage of the Cashmere cavern and looked up the narrow ventilation shaft receding to a small circular opening some 20 metres above me, I felt a shiver of fascination and amusement as I thought of my New Zealand colleagues being raised and lowered by a rope harness". It was Silverman who proposed the optical activity of an atomic gas as a consequence of earth rotation, but can it be detected a guarded affirmative answer brings us back to the cashmere cavern see chapter 9. [8] After the war the Cracroft family secured the property by bulldozed earth over the main tunnel entrance areas so isolating the adits, and filled in the cellar where the officers tunnel accessed the homestead and, concreted over a 60' ventilation shaft.

The ventilation shaft access method was clearly unsustainable for us in the long term. We gained permission to open one adit entrance, securing it with an electronic door. This greatly improved access. The project caught the popular imagination and was soon given a lot of publicity in the press and on national television.

Figure 23 opening west adit, Figure 24 Security door

we opened the west adit since we had found from inspection inside the cavern that the east adit has suffered from significant roof fall problems. The university added some wooden retaining walls and an electronic door with an electronic radio frequency controlled security system, one which has proved vital to combat vandalism. We wrote to the Canterbury Hospital Board for permission to use their property for access and were delight to get an affirmative response in June 1988. So I would like at this point to gladly acknowledge the immense help and understanding of the whole Cracroft family in permitting this whole project also Morrie Polson's, and Clive Rowe's, and Rob Thirkettle's, negotiating skills with the City Council Historic Places Trust Fire Services etc . All this has obviously has been vital to our project. Nowadays the ventilation shaft is the entrance point for power water telephone services and is taken simply for extreme emergency access. In the 23 years subsequently there has been no call to use that shaft or to return to our initial abseiling despite several major earthquakes in Christchurch in 2010-2012. The rock structure of the cavern is regularly monitored and we shall note the effects of these earthquake later.

We have had our share of medical emergencies

Cavern discoveries were not confined to matters of physics significance. On one occasion Morrie Poulton slipped on wet grass uphill behind the cavern entrance [we were in process of installing an antenna to link us with the University directly] and broke his ankle. As it was a remote area for help he staggered to the cavern door opening the electronic lock got himself to the rear of the cavern where he could phone us for help. We found a pretty sore and ashen Morrie stretched out in the cavern workshop he had built and rang for an ambulance whose crew had to use a special stretcher to get him out through the narrow access tunnel down the hill track and into hospital for an ankle pinning job.

In earlier days 1988 I collapsed walking up the hilly access track to the cavern with hospital ground staff to show them the cavern (the hospital owned the access land, we obviously had to maintain goodwill and approval on all sides and enormously appreciated it.) That hill track down to the Hospital access Roads was constructed from cavern spoil. That story ended some months later when I received a new aortic heart valve which has operated well since (although implicated in my 2001 stroke). Wybourne called that the first discovery of the ring laser program.

. Being in the cavern is not for most oppresive or claustrophobic (even if we could not see the sun we could learn a little of the weather outside within the cavern as the 'weather station' ultimately added to the data acquisition system (chapter 7) monitored the temperature pressure whose effects on the laser output was and important.

The C-I and the German C-II laser's were located in the best protected area of the cavern the 'military plotting room area' where the roof is particularly strong: the area of the following early photo.

Figure 25 scene at south end of major chamber from the officers tunnel on the day university technicians first descend the ventilation shaft

and arguably Professor Bilgers then made his most important discovery was making the aquaintance of Schreiber who introduced himself at one of Bilgers presentations on. C-I. Schreiber was already well established at the Wetzell station and later became professor at Munich Technical University.

Czechoslovakia

Bavarian foothills

Cham   Kötzel

Wettzell

Fundamental
station

Gottes-zell

Regensburg

Straubing   Deggen-dorf

Landshut

BAVARIA

Munich

Figure 26
1.   The above left map of Wetzell region in Bavarian foothills
2.   below Fundamental Station Wettzell operated by BKG Bundesamt für Kartographie und Geodäsie and MTU Munich Technical University programe Forschungseinrichtung Satellitengeodäsie, Sackenrieder Straße 25
D-93444 Kötzting Bavaria
(latitude: 49° 8' 41" N Longitude, 12° 52' 41" E)
the G laser is buried in the field beyond the radio telescope.

The location of this facility was determined in part by the suggestion that in the cold war days no aeroplanes flew over this area near the Czeck border and so there would be no hazard to pilots by pointing lasers directed at the moon, bouncing pulses of laser light off the reflectors left on the moon by the Apollo astronauts, and so commence a program to get milimetre precision in monitoring the earth-moon distance and monitoring its variation. Another endearing description of this delightful rural environment is the place "where the foxes and rabbits say good night to each other".[9]

Schreiber had earned a well deserved reputation as a Mr Fixit in this lunar ranging project. And was about to earn it again as regards the project to use ring lasers for significant geodetic measurements. An earlier such project by B Holling et al. using an argon ion laser had failed, in looking for alternative strategies Schreiber made the aquaintance of Hans Bilger.Schreiber visited Canterbury with Professor Bilger and Schreiber came again at least annually for 20 years. In effect Dr Schreiber joined the Canterbury ring laser project and has become a firm friend of the New Zealand group. His scientific skill has become legendary with us and has utterly transformed the project.

The BKG and Technical University of Munich programs combined satellite and lunar echo observations to determine very accurately the position and orientation of the earth relative to all celestial objects. So any new technique with the potential to give new information about how the reference frame of the earth was related to that of celestial objects is of interest. Indeed two such frame determinations are required.

The first step was to build a better-engineered 1 square metre square device C-II to improve on C-I.[11]

By the time of this opening event for C-II described in chapter 7 Dr Schreiber had laid a firm foundation for the German Dream for a far better laser 4 metres square which would have enhanced abilities and precision to monitor earth motion. C-I and C-II had proved the principle of the Canterbury approach as capable of this aim, but no one had built a 4 m × 4 m ring laser so the question outstanding was to test whether or not the designs could be scaled up. To this end Schreiber obtained money to build a simple prototype machine which was roughly 4 m square on the only optical table available to us of such a size namely the concrete walls built by the military to support the roof arches on one wall of the cashmere cavern near the adit entrance. This instrument (G0) was made to work with some difficulty. I initially made a calculation of mirror radii which did not ensure cavity stability (see chapter 2). I corrected the error almost immediately but the correction was not sufficiently clearly shared within the group. Once this was overcome G0 operated exactly as hoped and Schreiber now had clear evidence not only for the fabrication method proposed for G given the satisfactory operation of C-II but also that it would hit no fundamental impediment in scaling up the dimensions from C-I to meet the C-II design and achieve the German goal for G of 4 m × 4 m according to the recommendation originally set by Professor Bilger. So there was immense satisfaction in Germany and New Zealand when Professor Seeger President of the BKG not only opened the C-II facility(the plaque installed above the cavern for this occassion is reproduced in Chapter 7), but took the occasion of the opening of C-II as the time to announce that the (G) project had been given the go ahead by the Bundesampt. It was planned to install (G) in a purpose built cavern at the Fundamental Station Wetzell in the Bavarian Foothills described above.

Today the (G) laser instrument is the most stable and succesful accurate instrument in the world for the laboratory detection of earth rotation. This instrument has detected utterly novel physical Effects. See (Chapter 7.)As such it has fully realised the orginal dream of Professor Schenider. The Cashmere Laboratory has retained its useful role throughout as a useful testbed for new ideas. In particular it has demonstrated that one can build a ring lasers of still larger area. One ring laser (UG2) has been built encompassing the full area of the cavern viz a rectangle 40 m× 21 m however its results were not sufficiently stable as to increase knowledge of the earth rotation effects discovered with the lasers of intermediate size, in particular one (UG1) 21m × 17.5 m dimensions to which we returned when UG2 was closed down. Both of these extra large lasers used cross-adits between the main chambers left by the military in the cashmere cavern.

When Bill Wilsons aging mother died the (second) Cracroft family homestead was sold off and with it some adjoining land. The cavern area was made over to the City Council for recreational purposes as required of any new subdivision under Council laws. So the university had new owners to negotiate with and a 10 year lease has allowed access until the time of the earthquakes. Renegotiation and repairs have become necessary before the cavern can reopen. Like much of Christchurch we await a rebuild of a section before we can continue our geophysics research programme progress to date is as summarised in. Chapter 7 Like much of Christ church the cavern suffered badly in the major Christchurch earthquakes of 4 September, 2010 and an after-shock on 22 February 2011 (of magnitudes 7.1 and 6.3 R respectively). These quakes and particularly the second one demolished much of the business centre of Christchurch and badly affected many homes and caused 185 deaths. At the time of writing earthquake damage to the entrance tunnel has made the cavern out of bounds to all visitors. The remediation work involved would cost of the order of $100,000[12] and particularly to the roof near the workshop, shotcrete restoration on the walls of incoming adit (blue arrow) (and the bonding of the rocks at the north entrance red arrow).

Figure 27, Areas of the adit entrance needing earthquake repairs.

One condition of the lease was that the public should have access where possible and for some years the University has have operated these tours with help from the local Scouts Groups, more recently the city council parks unit has operated these through its Ranger section.

At one stage several 100' pine trees were felled on the west bank of the valley the thud each made in the cavern was significant. Morrie also enjoyed watching the opossums in the tree tops take their final ride down.

Overall scientific programme. Such projects are expensive. Funding has come initially from the physics department the University grants committee, from Germany BKG in supporting maintenance of C-II and the Technical University of Munich has essentially loaned us the $2M€ cost of the laser C-II German support also included the availability of Professor Schreiber.

The project has won 5 Marsden awards from the Royal Society of New Zealand, more than any other project at the University of Canterbury. In May 2014 the government announced six Centres of Research Excellence programs including its support for the ring laser program as part of the Dodd-Walls program in photonics including support for the ring-laser project with John-Paul Wells as a Principal Investigator[13]. This present cave closure shut-out unhappily affects the research program enormously so we are particularly anxious for the reinstatement of the cave.

The best documentary evidence and overview for the nature of the cavern as it was before the earthquakes is a television program which gives a vivid idea. It shows first the entrance drive to the cavern (the drive up the road built from cave spoil on to the flat area it leads to at the top, also built on excavated cave Spoil). The program continues with a walk around the cave interior showing the UG2 laser, this can be found on the net at:[14][15]

Endnotes

[1]   "The Port Hills of Christchurch." (Wellington: Read, 1978)
[2]   These notes above drawn in part from a city council parks unit publication on the Southern Group Headquarters in the cavern for Army Navy and Air Force during World War II
[3]   http://resources.ccc.govt.nz/files/Phase1ReportEuropeanCulturalHeritage-whataretheissues.pdf
[4]   Black Beech and Honeydew (Boston: Little, Brown 1965)
[5]   precisely a week after I was born.
[6]   L.J.Brown and J.H. Weeber, "Geology of the Christchurch urban area" p. 61

7   T.Williams and B. Chapman, "Underground tank N0.4, Christchurch" have a complete article in New Zealand Speleological Bulletin Vol.8,148, pp.211-216, Dec.1988.

8   **"And yet it moves: strange systems and subtle questions in physics."**
    (Cambridge UK: Cambridge 1993) p 211

9   http://www.fs.wettzell.de/
    http://www.bkg.bund.de/SharedDocs/Download/EN-BrochFly/BKG-Geodetic-Observatory-Wettzell-Brochure-EN,templateId=raw,property=publicationFile.pdf/BKG-Geodetic-Observatory-Wettzell-Brochure-EN.pdf

10 http://www.fesg.bv.tum.de/fesg.html

11 Wikipedia http://en.wikipedia.org/wiki/Cracroft_Caverns

12 while substantial I compare this with the 7 or more houses demolished in my street after the quakes

13 http://www.tec.govt.nz/About-us/News/Media-releases/Research-Excellence-Centres-Impressive-/

14 http://www.ecasttv.co.nz/program_detail.php?program_id=1921&channel_id=84&group_id=

15 An unusually wet winter in 2014 aided of course by earthquake history arguably caused a major rock fall of one roof section which demolished UG3 and puts the future of the cavern laboratory in doubt, but as of Nov. 2015 discussions between all parties are remarkably hopeful for the cavern's eventual restoration.

4 THE CANTERBURY RING LASER –ITS ORIGINS

The story I tell in this book includes some bends and twists that might be unexpected. The one discussed in this chapter chronologically the first could be ranked as the strangest.

The two reasons for discussing this topic here are it's relevance to our understanding of the Sagnac effect (Chapter 2) and to explain how I got into considering the significance of ring lasers experiments.

First I would like to discuss the issue of Clock Synchronization in polygonal interferometer. I have long had an interest in defending relativity, for example I replied to a spurious suggestion of Dingle.[1]

As a student, I once suffered from what I suppose can only be described about as typical student arrogance. One of my lecturers, Archie Ross (chapter 8) had given us what was to me a majestic course on operational relativity. A sequence of thought experiments starting from fundamentals that could be used to the principles of special relativity with its non classical conclusions over time dilation and length contraction twin paradox could all be derived logically from the premise that the speed of light was a constant. On being presented with this, it seemed to me that a natural question to ask was how this formulation could be modified to cope with a speed of light which depended on direction, whether or not it did so vary would be a question that would ultimately have to be decided by experiment interpreted within a suitably adjusted theoretical framework to a test of the matter. It was not seriously envisaged that the world was anisotropic in this manner but merely that such a proposal was imaginable and therefore worth exploring and testing. But one could only hope to do that by investigating their logical consequences in the formulation of relativity. This was years before the experimental ring-laser projects recorded in this book were contemplated. It was many years before I realised that this was not an original idea and that others had thought it worth exploring theoretically and experimentally or indeed how best to go about introducing such an innovation. I presented Archie with an attempt at such a formulation of an anisotropic form of operational relativity but must confess I was too inexperienced to have seen this through as logically and completely as one might have. Later as a PhD student in London I picked up a philosophy journal and came across an excellent paper by Winnie[2] which answered the very question I had attempted and in the process had corrected other relevant work. This matter nagged me for years and I interested two students Ron Anderson and Kumar Vetheraniam in perusing it as a project. The history of the Canterbury ring laser projects really starts from that point. That Ron and I generalised the choice of synchrony to and arbitrary position dependent field $\kappa(x)$ which had to be solenoid-al[3] to preserve the equality of light speeds in both senses around a polygon we give more details later in this Chapter. relevant references can be found from my initial papers with Ron[4][5][6] Ron and I with my first experimental collaborator Hans Bilger wrote up our own centenary contribution on the Sagnac effect in American Journal of Physics[7]

Ultimately Ron, Kumar and I wrote up a magnum opus on the topic of synchrony in special relativity[8] In the meantime Ron had pursued this interest as a lecturer in philosophy at Boston College, We had grander plans for further such work but these crumbled at Ron Anderson's untimely death in June 2007[9]. All of this added to my long term interest in relativity

Einstein was convinced by Maxwell's theory of electromagnetism that the speed of light is best regarded a universal constant in developing relativity given an operational scheme for setting up a system of clocks through an inertial frame with which to conduct thought experiments. A remote clock could be standardised against a master clock at the origin by sending a light pulse between then and then back again. From the remote clock. Then if one assumes that both pulses travelled at the same speed then minimal adjustment is needed to the remote clock to allow for the travel distance. One can check that adjustment by repeating this experiment as often as necessary to correct any initial maladjustment of the remote clock.

A full account of the issues is given in the above mentioned review. The operational approach to relativity can clearly be reworked on the assumption of equations (1,2 below) rather than the conventional assumption ε= ½ as Poincare pointed out, conventions are necessary but not arbitrary, see Max Jammer's work 'the concept of simultaneity' debate and misunderstanding continues Selerri has also objected to the Anderson et al choice of synchronization supposedly on the grounds that it contradicted relativity. His objections have been examined by Rizzi Ruggiero.

To explain anisotropic relativity further. First let us suppose as in the below discussion that the speed of light is different in opposite directions. Since round trip speed of light is found to be the constant c from experiment (see the chapter 5) this requires a link between these two speeds discussed here, one

account is given in[11]. If the speed from A to B is

$$c/2\varepsilon \qquad\qquad (1)$$

then the speed of light B to A must therefore be

$$c/2(1-\varepsilon) \qquad\qquad (2)$$

these formulae keeps the round trip A to B to A speed fixed for any value of ε. Nearly all texts of relativity simply choose these speeds are equal, so that we must then have ε= ½ so equations (1,2) have become the natural choice for the conventionality of the speed of light.

The next figure depicts such a non standard choice when synchronising a master M and remote clock N, in which the light speeds are taken to be v and 2v corresponding to equations 1 and 2.

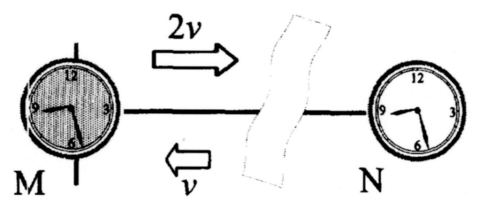

Figure 28, Clock synchronisation in relativity in Einstein synchrony ($\varepsilon=½$) light flashes are sent from a master clock to a remote clock and back thus synchronising clock N with M assuming an isotropic speed of light c. if that is not accepted assumption (*2v* $\neq v$) so $\varepsilon\neq$ ½ , remote clocks N must be adjusted accordingly by an amount depending on ε.

The choice $\varepsilon=$ ½ for this guarantees the basic observable of the round trip speed of light c is a constant in agreement with relativity. Any other choice of ε corresponds to a deliberate re-setting of the synchronization of the clock distant from the origin and master clock. Because the equations 1,2 of this chapter this will preserve the constancy of the round trip speed of light. There are some consequences of this anisotropic operational scheme. All relative velocities are dependent on ε. In addition the length contraction and time dilation factors are also ε dependent. in fact this gives them a lineal dependence on velocity not just the well known surd. So that time dilation exists even for clocks which are infinitely slow moving, there is a cumulative effect in their final motion. We outline the explanation below why when such clocks are used to synchronise the clocks in an inertial frame, the result is that a network synchronised by slow clock transport turns out actually to satisfy the Einstein synchrony convention $\varepsilon=$ ½. so that one cannot detect an anisotropic speed of light to disproving nor can one use slow clock transport.

One might hope that any anisotropy in the speed of light would be manifested by interference experiments as the Michelson Morley experiment in which light waves are beaten against light waves. This hope is also false. So the counter-intuitive fact is that no other value of ε than ½ cannot show up in any experiment. Any other choice for ε simply amounts to a non-standard choice for synchronization of clocks in an inertial frame of reference. This ε dependence being imposed here for all that it seems arbitrary is purely a change of the definition of coordinate time, after all millions of air-line passengers from New Zealand to the USA have flown across the International Date Line without suffering ill effects of the international date line, which therefore becomes devoid of biological significance (nor does this imply any central importance in science any more than does the choice of Greenwich over and above Paris for the zero of longitude.)

The at first sight surprising conclusion that ε cannot be measured, needs some detailed defence. Many people think they have found relativity can be a way to measure ε until you really look into the details of their method. It is one of those not always particularly welcome hobbies of mine to look for flaws in proposal to measure ε, but there are some things in science that someone has to do. Unfortunately in the history of relativistic physics here has been badly misrepresented by both some physicists (even some well known committed relativists! have fallen into traps even Clifford Wills did this[12]) as well as many philosophers of science. Mercifully the need to keep setting the record straight is not as acute just now as it was even a few years ago. A full account of the issues that have been debated would requires a full book some help may be found in.[13] In the operational approach (the chapter on relativity illustrates how the conclusions of operation relativity maybe reworked on the assumption of equations (1,2 above)a detailed presentation is in Winnie[14] and the results are quite different to those of standard relativity, relative velocities time dilation and length contraction are all ε-dependent.

With this insight one may look critically at the historic means of determining the speeds of light e.g. Ole Rømer's 1644-1710 studied of seasonal changes in the apparent position of the moons of Jupiter indicated in the right hand picture below, the apparent variation in their positions can only be understood by recognising that the light travel time from Jupiter changes with its distance from the sun. Laboratory measurements also can not be devised to yield an apparently one-way speed of light. For example One may use of rotating toothed wheels while retro-reflecting light on a return path from one to the other on the argument that light travel times will inevitable cause the beam to be blocked by one set of teeth on the return passage of the light if the pair of wheels are rotated sufficiently quickly. One can certainly get the standard speed of light this way but.in the generalised operational relativity mentioned above in which an ε-dependent speed of light is postulated, the possibility of a truly one-way measurements is denied by the fact that the synchronisation of the two rotating wheels so that relative timing is ε-dependent. The same fact denies an attempt using rotating prismatic mirrors as depicted below these also fail to give any value for ε all such experiments except ε= ½. The same problem frustrates Rømer's method of measuring the speed of light in the sense that it is not a one-way measurement, despite the enormous distance between Jupiter and the Earth, since the clocks formed by the Moons of Jupiter in the periods of their revolutions and the clock on Earth differ if ε is not its standard value.

Figure 29, Old laboratory measurements of the speed of light cannot reveal any anisotropy in it, nor can Rømers method for determining the speed of light by observing the moons of Jupiter observe anisotropy in light speed.

A plausible method of determining an absolute synchronisation of clocks, is that of slow clock transport which would appear to settle all arguments on relating a master clock to others within an inertial frame. In the figure below the upper set of clocks are Einstein synchronised, i.e. We assume isotropy for c ε=½. In the middle the clocks are slowly transported slowly to define remote synchronisation again assuming isotropy they therefore agree with the clocks in the top row. But there is a conflict between these and slow transport synchronise clocks depicted in the middle and those at the bottom row for which an anisotropic ε-dependant speed of light is assumed. So it is clear that different methods of synchronising clocks can give genuinely different results (a point sufficiently well illustrated just by changing ε). For this situation it seems difficult to put ones finger on where the conflict arise, the basic problem is because if the speed of light were anisotropic the effects of that must be logically followed through in making such comparisons between the last two rows of clocks and it is not logical to say that ε=½ in the second row, in fact no ε value has been assumed. In this case to achieve full consistency between the last two rows one has to allow that if for the time dilation factor appropriate for the middle clocks which is ε-dependent and that this is significant for the middle set of clocks even when they are moved infinitely slowly, for details of the proof of consistency resolving this issue see the paper by Winnie. This time dilation effect arises and is of importance for any non-standard values of ε.

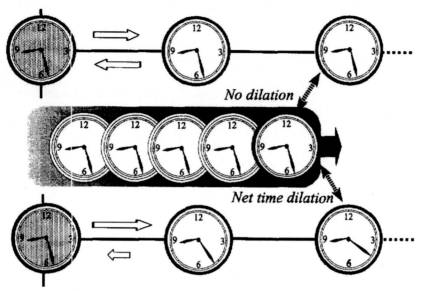

Figure 30, Synchronisation by slow clock transport

The long and short of all this discussion is that any posited one-way speed of light other than c proves to be invisible despite the most ingenious attempts one may make to reveal it.

Ohanian[15] objects against ε-dependent synchrony that it introduces fictitious forces, and so compromises F=ma. I don't know that this objection has ever been urged against Newtonian mechanics itself, with its fictitious centripetal forces and coreolis forces. Ohanians objection has been countermanded by others, and this is not the place for a full study, but I mention the names Brown[16] Kalauber MacDonald as authors contributing usefully to this debate this misunderstanding continues. In fact the conventionality of synchronisation was recognised very early in the history of relativity theory. Similarly contentious issues relating to synchronisation have been invented and adducted in order to denigrate and obfuscate the standard descriptions of the Sagnac effect as explained in chapter 2.

The above serves to illustrate the point that questions about how the operational methods of relativity would be affected if one introduced anisotropy of the speed of light. In fact it turned out that although Einstein's and Reickenbacks approach to the conventionality of synchrony was adequate and unequivocal. The most fundamental clarification I know of the operational approach is Brown's because of the quite exceptional care and erudition he shows in discussing many apparently insignificant points with respect to such things as the assumption of reciprocity in[17].

Relativists accept there can be no real dispute over matters like the conventionality of synchronisation or any other aspect of the Sagnac effect General Relativity has removed all basis for any dispute. As mentioned above and in Chapter 2 there are exceptional presentations even today. One can extend equations 1,2 of this Chapter to describe a totally anisotropic, and fully variable choice of synchronisation t=t-κ.x, where the field (ε(x) if for convenience we replaced this field by κ(x)=2 (ε(x) -1) in a position-dependent field the field κ field must be irrotational[18]viz $\nabla \times \kappa(x) = 0$. Then As fully studies in relativity and our experiments require the Sagnac effect is to be guaranteed[19] in any polygon.

Polygons

We now consider effects in the more complicated (circular, square polygon) geometries characteristic of ring lasers. Given a circle or polygon) A B .C .D .A what experiment permits us to conclude that the speed of light is the same in each direction. At one corner of the polygon simple answer is to have the master clock, it can measure both speeds. Reichenbach calls this the round trip axiom: If from a point A of the static system two light signals are sent in opposite directions along a closed triangular path ABCA, they will return simultaneously".

Reichenbach attributes the recognition of this axiom to Einstein. Weyl has a similar axiom, which he refers to as an Erfirungstatsuche ('fact of experience.')

Weyl's axiom includes Reichenbach's axiom. These axioms are connected by, while distinguishable from, the "light postulate" (that the two-way speed of light is isotropic).

(An alternative interpretation of Weyl's Erfahrungstatsache makes it more akin to Robertson's approach by denying the light postulate until confirmed by the 'experience' of the Michelson-Morley experiment.) Within general relativity, the Reichenbach round-trip axiom forms a condition on the definition of a local inertial frame; it must not only be in free fall but also be non-rotating in this sense, so as to have no Sagnac effect[20].

$$c\left(\overleftarrow{ABCD\_A}\right) = c\left(\overrightarrow{ADCB\_A}\right)$$ Early in the history of relativity this was recognised, as a 'fact of experience' i.e. a matter dictated by experiment. A number of relevant experiments are listed in a correspondence with Erlickson as mentioned below see also[21].

 Since this is an unusual question in relativity I wish to explain my reason for fuller confidence in this belief viz. that the speed of light each way around a polygonal path is the same I shall do so historically using the particular examples which forced me to face each of these questions seriously.

In recent years it has been popular to dispute the issue (which goes to the centre of the Sagnac effect, chapter 2.)

The correspondence was triggered by a paper published by Windberger and Mossell [22]. where they suggested there was some fundamental interest in studying a polygonal In which Light flowed from a sources A via each to mirrors B, C, to think of a detector D. the authors thought they could distinguish the speeds of light in different directions but relativity has taught us that the measured speed of light is always c. and the authors were not aware of the manner in which the conventionality of simultaneity upset their conclusions.

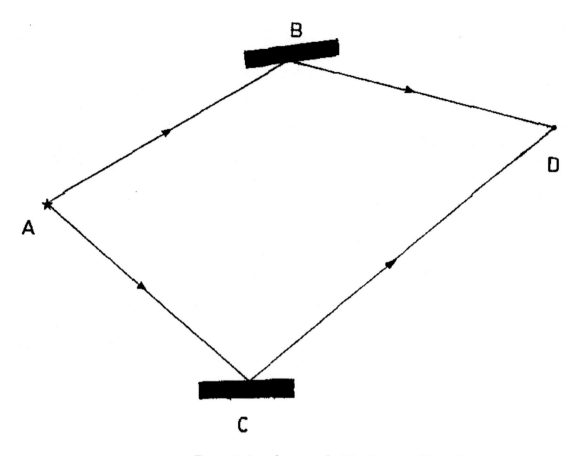

Figure 31, Interferometer for Windberger and Mossell

The. Argument was that the geometry was sufficiently different on different paths. However even if one looks at this situation in an arbitrary, choice of synchronisation a definitive conclusion is possible. Let us consider the various transit times for a light beam. One is therefore considering four possible loops

ABDBA ABDCA ACDBA ACDA

Let the transit times of a light beam around these paths

ABDBA	t_1
DBDCA	t_2
ACDBA	t_3
DCDCA	t_4

I remarked in 1973 when writing this that rotations varied these times and that the distances and light velocities refer to their conventional values measured on round trip measurements.in operational relativity.

The distances are measures by light travel time . These references to rotation were made long before I got involved with ring lasers. And make no presumptions on the effect of rotations on light travel times, however the fact is that Doppler effects at mirror reflection show there is an effect when the system is rotated, as this is one way of understanding the generation of the Sagnac effect.[23] See also the consideration of Doppler shifts at a moving mirror though this is not immediately relevant [24].To this extent my 1973 analysis was incomplete

These times will correspond to the round trip distances divided by the invariable speed of light.

the total round trip travel times $t_1 + t_2 + t_3 + t_4$ are all equal since one clock, at A set up under Einstein synchronisation (which as we have seen cannot be disproved by other experiments) suffices to measure them. Whatever synchronisation scheme is adopted and they are equal to since otherwise here would be an apparent discrepancy between the speeds of light in opposite ways around a polygon violating the 'fact of experience' above.

the value $(AD+BD+AC+CD)/c$

and any difference in transit times will create over a shift in an interferogram

$$\phi = 2\pi\nu\,(t_1 - t_2)$$

$\phi = 2\pi\nu\,(AB+BD-AC-CD)/c$

These relations showed to me in 1971 that there was no fundamental interest in the of Weinberger and Mossel configuration, this was years before I was involved with ring lasers. I wrote in 1971 that these relationships with be unchanged by rotations, e.g. by the Sagnac frequency shift as indeed other effects of rotation which were not an interest of Weinberger and Mosel. So the conclusions of the Weinberger Mosel experiment – a measurement of ϕ was not an open question[25].

I was challenged by Erlichson.[26] On the irrelevant grounds of the assumption of ether theory in a parallelogram interferometer. This demanded a reply which I offered on the following lines. "Is the Apparent Speed of Light Independent of the Sense in Which It Traverses a Closed Polygonal Path?"[27] I listed a number of experiments in which polygonal interferometers had never been bedevilled, by changes in light speed, whether producing diurnally or the earth rotated through the ether or otherwise. These included the 1851 Fizeau interferometer where a light beam is sent in part through water before being made to interfere with itself, the 1913 Michelson interferometer where a light beam is split one beam being sent on a horizontal path and he other on a vertical one before they are reflected back and made to interfer, the 1925 Michelson and Gale interferometer to detect earth rotation from its Sagnac effect in a large interferometer (see chapter 7), Holography which depends on optical interference experiments in a parallelogram type interferometer. And I then added that the best evidence of an equality of light speeds in opposite directions around a polygon was that gained from ring lasers. I never foresaw the momentous consequences of that statement of mine for my own future culminating in the projects described in this book. So in summary while philosophy and physics were infected by totally wrong ideas on the validity or otherwise of alternative synchronisation schemes all these years after Einstein. fortunately problems are much less obvious in the literature of science than they once were, so, it is fair to say on the one hand that the recognition of the truth of the matter is much clearer now than it used to be. Unfortunately it also has to be said that errors in much discussion of related topics still severely bedevil the very basic questions of interpretation of the Sagnac effect whose nature is vital to the operation of ring lasers (Chapter 2). Hence there is no alternative in a book of this type to making the many related matters clear as possible.

The appearance of one very confused book Rizzi and Ruggieri. many give the impression that the existence of the Sagnac effect reveals a fundamental conflict with relativity. The problems with this book are illustrated in a review by Tucker. this is discussed in detail in chapter 2.

It was many years after publishing this reply to Erlickson: that I received the following letter out of the blue from Professor. Hans R Bilger, then at Oklahoma State University Stillwater. From this the Canterbury ring laser project was born. Hans wrote as under more details of this letter are quoted in chapter 8.

"13 January 1982, Dear Dr Stedman I recently came across an old letter of yours[28] where you discussed optical closed path devices. What are your thoughts about the general relativist analogue to the Arhananov Bohm effect (quote from your AJP Letter), the Ring Laser can in principle be made large, of the order of m² at least though. Nobody has yet claimed to have reached the quantum limit in such devices(industry has pushed the state of the art in devices of order (dm)² up to now. Do you think about the possible to find a gravitational AB effect due to the as the moon and/or the sun. This has the advantage of furnishing variable acceleration vector directions at the ring laser site. In case these questions still meet your present interest I would greatly appreciate your thoughts on the subject.

When light beams are sent both ways around a closed interferometer they can be beaten together the beating produces interference fringes give extremely sensitive detection of the rotation of the whole device. This is the so called Sagnac effect first detected in 1913. (GES Its study will be a principal concern of this work. Bilger was much more aware than I was. That even at the time I wrote those words enormous strides were being taken in improvement of mirror quality and the devising of devices to measure rotation for inertial navigation the chapter 2 explains the central ideas.) Including the use multilayer mirrors with SiO_2 and TiO_2 alternating up to 22 layers which gives previously unheard of reflection efficiencies of 99.99999% or better and with them a dramatic improvement in cavity quality factors and the associated reduction in noise meant that devices with a perimeter of order 60 cm would soon become standard for measuring the roll pitch and yaw of all kinds of aeroplane, missiles and such avionic devices, these were based on He-Ne lasers with DC excitation, this technology matured and became standard avionics for jet aeroplanes in the mid 1980's along with accelerometers they provided the key sensors in inertial navigation systems, details were essentially military secrets, and it was the 1970's before commercial purchase of super mirror technology became possible. When our program commenced were able to prevail on the generosity of one mirror manufacturer to give us some mirrors for our very non military research project. As far as Erikson's objections are concerned.

The essential point here is that if the speed of light varied with the sense of propagation in a polygon the effect would have been immediately visible in any of these devices as a frequency shift in the interferogram obtained on beating the counter propagating light beams and that with extreme sensitivity. Hans explained that part of his interest was to have three good lasers around the world at different latitudes, in say Europe, New Zealand then one at Nairobi on the equator for example.

Hans R Bilger"

This started a furious correspondence between us on a multitude of things these were in the days before e-mail existed so that I kept the universities one and only fax machine busy! (when one of Hans first statements on arriving at Canterbury was that we should buy a computer for the project I internally blanched! I was not yet living in that kind of world once in it I have never been able to get out of it and that was the least of our technical problems.) At this point it should at last be clear why for the reasons indicated through this Chapter quite early in my career I got involved in discussing the question, how do we know that the speed of light around a ring is the same in both directions, clock-wise and anti-clock wise surely that at least is amenable to experimental test and can be no mere convention, only one clock is needed to time both directions. This also required an understanding of the Sagnac effect.

Endnotes

[1]   G. E. Stedman Nature **244** 27 (1973)

[2]   J.A.Winnie Phil. Science 37 81,223 (1970)

[3]   this means that if $\kappa(x)$ were regarded as the velocity field of a fluid and if a little paddlewheel were dipped in it should not rotate, mathematically $\nabla \times \kappa = 0$.

[4]   Ronald Anderson and Geoffrey E Stedman, Distance and the conventionality of simultaneity in special relativity Found. Phys. Lett. 5 199--220 (1992).

[5]   Ronald Anderson and Geoffrey E Stedman, Distance and the conventionality of simultaneity in special relativity Found. Phys. Lett. **5** 199--220 (1992).

[6]   reply to Coleman and Korté Found. Phys. Lett. **7**, 273--283 (1994).

[7]   GEN B

[8]   ...).GEN G

[9]   http://www.bc.edu/bc_org/rvp/pubaf/07/andersonobit.html

[10]   Indeed Ohanian is on record for objecting to Einstein's statements on synchronizations. Physics Today April 2006 p 10 "Einstein overlooked the validity of Newton's laws at low speeds, permits the use of simple mechanical methods of synchronization such as slow clock transport or sound signals. Here Ohanian however ignores the effect of time dilation on slow clock transport within all synchronization schemes other than the one he favours.

[11]   GEN G equations 1,2 below correct the expressions in this reference and preserve consistency over the choice of +x axis

[12]   C.M. Will, Phys. Rev. D **45** (1992) 403

[13]   GEN G.

[14]   J.A.Winnie Phil. Science **37** 81,223 (1970)

[15]   H Ohanian American J Phys**72** 141-148 (2004);

[16]   GEN H.

[17]   General reference G

[18]   a vector field representing velocity of a fluid with velocity distribution $\kappa(x)$such that a paddlewheel was immersed in it would tend to revolve.

[19]   Ronald Anderson and Geoffrey E Stedman, Distance and the conventionality of simultaneity in special relativity Found. Phys. Lett. **5** 199--220 (1992).

[20]   GEN G

[21]   GEN G.

[22]   Weinberger and Mossell, American Journal of Physics, **39** 606-609 (1971)

[23]   Dresden M and Yang C N 1979 Phys. Rev. D **20** 1846–8

[24]   earth rotation via Doppler shifts? Geoffrey E. Stedman Am. J. Phys. 75 778 (2007)

[25]   G. E. Stedman Amer. J. Phys. 40 782-784 (1972)

[26]   Erlickson American Journal of Physics 41 1298-9 1973

[27]   Reply to Erlickson: G. E. Stedman Am. J. Phys. 41 1300-1302 (1973)

[28]   Reply to Erlickson: G. E. Stedman Am. J. Phys. 41 1300-1302 (1973)

5 Relativity

Space Time Physics

Every one has heard of the fourth dimension. It is time. Along with space's three dimensions, up/down sideways, each way north/south, east/west, these are the framework, a grid used to discuss any events that happen through the world. Every thing that has ever happened can be given its unique location by specifying its position within space-time. Relativity has shown that there are some unexpected connections and restrictions between the time numbers that are assigned to events and the space numbers. These go to the roots of how we define space and time, and are described using formidable technical mathematics that will not be touched on in this book, an incisive and well informed study of many foundational ideas is that of Harvey Brown.[1]

The Principle Of Relativity

Quite simply the notion of relativity is that only relative motion is detectable. In short, the absolute speed of anything cannot be measured. All the basic theories of physics are consistent with this, and all experiments devised to date only confirm it. The matter can be rephrased this way, can passengers in a train or a plane do any experiment which gives their absolute speed through space? Indeed, if the vehicle is moving when moving smoothly and comfortably how can we be sure that we are really moving? There are several levels at which this question can be discussed.

How can You Measure The Absolute Speed Of A Vehicle With Respect To The Surface Of The Earth?

Yes this is a relative speed and is not forbidden. As a train passenger looks through the windows of a train, the telegraph posts going past would confirm that. The plane passenger might look at the clouds or land below or the stars moving above, which are also giveaways. The pilot will of course know because the aircraft navigation system will display an air speed of say 805 km per hour or so through the air, for that speed can be measured. The standard way for an airplane to measuring this air speed involves a pitot tube with its two ends one open to the airflow measuring a pressure between its ends. For the pilot this is backed up by GPS, inertial navigation systems, radio communications with the control towers, navigation beacons, and of course is what he can see through his own windows. The Inertial guidance systems are used nowadays for larger planes. will have the accelerometers in the cockpit measuring linear acceleration in the $x\,y\,z$ directions, integrated over time these must give his airspeed. That is indeed a central part of the guidance system, (is to measure the rotation of the aircraft about three perpendicular axes, the roll, the pitch and the yaw rates with ring laser gyros, and integrate over time those readings and permit a pilot to automate aeroplane landings even after intercontinental flights). The passenger could work out his airspeed too, without windows (his mates may want to watch the movie, say) or pitot tubes. If he has say a duplicate of the pilot's inertial guidance system in the cabin with him, he need not communicate with the outside world at all to detect that relative motion.

It Is A Different Thing However When It Comes To Measuring One's Absolute Speed Of The Plane

The speed through space of the earth surface, is 2,688 km per hour because of earth's rotation alone, the speed at which the earth is moving round the sun is 107,826 km /h. The only reason astronomers can measure these at all is because they can compare the earth's motion with the sun's. No standalone device with the passenger or pilot can measure this speed and show it on its dials. Of course a good astronomer on board with the right tools might use a good telescope and make reference to the sun's motion and deduce these speeds. But it is always a relative motion only that can be detected. Do not imagine the issue here is a simple one. But even so these instruments do not provide the absolute speed in through space. The principle of relativity says amongst other things that such an experiment is impossible, and that only a relative motion is measurable. The passenger or pilot constrained to ignore windows and telescopes is condemned to ignorance of such speeds and ultimately to ignorance of our absolute speed through space. If one cannot do that, what objective scientific meaning is there to the idea of absolute motion through space? no apparatus can be built based on orthodox science that would detect in isolation absolute motion. One has to think of every possible type of experiment without exception before making such a strong statement. There are obviously two categories to consider—mechanical experiment and optical experiments. We have to consider these cases separately.

Newton's Mechanics And Relativity

Isaac Newton in the 1600's founded his mechanics on the assumption that the stars defined an absolute inertial rest frame. True, some stars move in our sky and the motion is observable over time, but for all that the average position of stars was at absolute rest for Newton, and was a reference for the motion of every other object. Only in that way could he make it a law that the acceleration of any object with respect to the stars required a force proportional to the mass of the object. To be sure at night and even in an aeroplane we see the heavenly bodies rotating in the sky 15° per hour, because of earth rotation. That is to Newton no evidence of our absolute motion, since the stars are our reference for that. What clinches the principle of relativity for Newtonian mechanics is that if one takes Newton's equations and adds a constant speed to everything, v say, (i.e. change our 'absolute speed') the constant v simply disappears and the equations have the form they had before the constant v was introduced. So the conclusion is that the absolute speed of anything will never be visible in any mechanics experiment. A simple illustration is that when playing tennis on board a ship, you do not have to change one's tennis technique because of the ship's speed through smooth water the ships speed does not enter the equations that Newton would write down for the ball or the racquet. It is true that I am oversimplifying here. Naval gunners soon found that the aiming of cannons demanded an adjustment for the 'Coriolis forces' associated with earth rotation, so tennis players near the poles in principle have to compensate the earth's rotation by changing their playing technique. The change is minute, and of no consequence for tennis games. A physicist in recent Centuries would conclude: "relativity holds true, in mechanics."

Optics

During the late 1800's in particular, it was realised that an interesting but unanswered question was: Does the same principle of relativity—the un-testability of any absolute speed through space—hold true for any optical experiments also? The key equations of optics were finally discovered almost through accident when Maxwell was investigating electricity and magnetism in 1867, and they implied the result that the answer was *yes*, and that the speed of light was an ultimate in today's units c= 299,792,458 m/s, was the maximum permitted by the theories of physics.

For the reason was the same as in mechanics, a constant speed v added to everything had no effect on the form of his equations of optics, so that and absolute speed was again then undetectable. Thus, relativity still held true even in optics, and a constant speed was still invisible. This had been an open question in science. That did not stop people trying to detect an absolute speed v because by now it was realised that optics experiments were capable of enormous accuracy since light was now known to be a wave and the effect of adding different waves together to see them 'beating' allowed an Extraordinary close scrutiny of the effects of any motion. Students are nowadays taught very elegant mathematical ways of understanding this, thanks to relativity. Einstein himself had a very neat way of illustrating the principle of relativity in electromagnetism in the dynamo example below. That means in particular that no optics experiment in the aircraft cabin could ever measure the plane's absolute speed. This is contrary to our classical common sense, by classical I mean in a Newtonian tradition. If you walk down an aircraft cabin carrying a torch one would normally-assume that the light speed was increased over the speed of the plane by your walking speed, but if absolute speed is not measurable, an experiment to measure the speed of light, which could easily be performed inside an aircraft, would not reveal the aircraft speed.

This was such an important matter that it was not considered closed until careful experiments were done to check. These were done to see if earth rotation for example affected the speed of light, so as to see if people could find the absolute speed of the earth from a single experiment.

These were first performed on the ground during the late 1800's and showed convincingly what the equations said, that there was no laboratory experiment not even an optical one that could show up the observers absolute speed travelling through space. So during the early 19th Century a physicist would still have said that an experiment with light should give an absolute speed. Light travels at a definite speed within a vacuum—in today's units this is 299 792 458 m/s a statement that is exact by definition. All this raises another question, with respect to which object or material is the light moving at this speed? Now light is unique—sound needs air but light needs nothing to come to us even through the barrenness of space from the stars to earth. This seemed puzzling to some 19th century scientists, who worried by this, tried to think of strange mechanical models of the vacuum, imagining light to be connected to some material or mechanical system of gear wheels etc., to sustain its travels. This elusive material was called the ether.

During the late 19th century, many people struggled to do experiments with light particularly, that might possibly show such absolute motion, like the absolute speed of a train or laboratory. All of these experiments failed. The motion of the source or medium did not affect the speed of light, not even in starlight from orbiting binary stars. It rather appeared that the principle of relativity could be extended beyond mechanics to prove valid for optics also. All of this was vindicated fully through the work of. James Clerk Maxwell, whose theory of light has been justified fully has spawned Radio, TV and many other things directly. He was following the remarkable insights of Michael Faraday who with others first understood experiments when electric and magnetic fields changed with time. Maxwell's theory is an enduring monument to the power of theory in science and showed beyond doubt that light was a wave motion of electric and magnetic fields. The basic ideas embedded within Maxwell's theory can be stated very simply. In the 1830's Oersted and Ampere had realised that if changing the position of electric charges (so causing electric currents) created magnetic fields (which make compass needles swing) magnetic fields thus arise from moving electric fields. Conversely, one of Faraday's greatest discoveries was the dynamo, in which a loop of wire rotating in a magnetic field creates an electric current flowing in the wire. The electric current created by moving a wire in a magnetic field is the principle behind major electricity generators in the world today, and it was Faraday who discovered it. Maxwell put both these ideas in to concise mathematical form. That form invited the question: could one consider the possibility of a combination effect, one in which a moving electric field could move in such a way as to create a magnetic field whose motion exactly produced the electric field first postulated?

Moving magnetic field $\Leftrightarrow$ Moving electric field The motion of each in space and time would no longer need magnets, dynamos, charges or wires, but each field supporting the other when oscillating within time and space. The result was a mathematical picture of a light beam, as both fields, to support their mutual driving motion, have to travel through space at fixed speed given purely from the theory from Maxwell's laws of electricity and magnetism (in particular Maxwell made an innovative modification of the equations describing the production of magnetic fields)'. This speed turned out to be $c = 1/\sqrt{\varepsilon\mu}$ where ε is the electric (permittivity) constant of free space (itself expressing the ease of creating electric fields from charges in space) and μ the corresponding magnetic (permeability) constant (itself expressing the ease with which from magnetic fields can be created in space). (There is a very beautiful symmetry between μ ε in the Maxwell equations you can see the balance between magnetic and electric effects, in the formulae for the speed of light $c = \dfrac{1}{\sqrt{\varepsilon\mu}}$ this understanding was fundamental to the new understanding of electromagnetism). The wonderful thing was that when Maxwell put the then-known values of these constants μ, ε into his equation, it came out to $c \sim 310\,740\,000$ km/s from the calculation, close to the experimental value of the speed of light 299,792,458 m/s, now for the first time correctly predicted by Maxwell from a solid physical theory. This is the type of discovery that any scientist dreams of, and is undoubtedly one of the very greatest scientific discoveries of the last several hundred years. Maxwell had shown that light was an electrodynamics wave, a combined motion of electric and magnetic fields. He was using a small extension of what had been proved experimental to come to the idea. Moving magnetic field ⇔ Moving electric field.

As mentioned one consequence of Maxwell's theory is the fundamental idea of relativity that absolute motion cannot be detected, even when it came to light beams.

Dynamo

Indeed, Einstein begins his first famous paper on relativity by a simple example of this. He says that the principle of relativity holds for Faraday's dynamo effect: You can move a wire near a magnet to create an electric current, a voltage is produced; on the other hand if a magnet is moved near the stationary wire so that the relative motion is the same, the same voltage is induced. And experiment confirmed that this shows that no 'Absolute motion' of the wire magnet system affects the voltages. So the principle of relativity holds for every prediction of Maxwell's theory, not only this one, so the principle of relativity thus holds for optics too. Einstein started with the idea that no optical measurement can detect our absolute speed, and that the speed of light is a fixed constant of nature, in particular the speed of light does not change even when changing the speed of the source. The speed of light is constant to all steadily moving observer's. This is a radical statement with immediate and radical consequences for our understanding of space-time. In the remainder of this chapter we want to mention some of its logical consequences. It has been convincingly verified in all experimental tests. All the experiments of the 19th and 20th centuries force agreement with this statement. The most famous experiment was the 1887 Michelson-Morley experiment. In effect it measured the difference between to-and-fro light speeds in two directions at right angles. This experiment was certainly precise enough to show that if the idea of the ether was right, one should find a changing speed of light depended on direction or on the season of the year or speed of the earth through the ether. It was just one of many, and has been repeated many times with far more precision in the 20th century, but the other types of experiments have added conclusive evidence at the critical time when classical ideas of light speed are simply inadequate. Since that day, many far more accurate experiments have been done. For example, Brillet and Hall[2] did a more sensitive test of the using two stabilised lasers with an enhancement in accuracy over the Michelson-Morley experiment by a large factor of only confirming its result. There is again no evidence in favour of the ether's existence as a supposed medium to carry light waves.

A Brief Summary Of Some Curious Predictions Of Relativity

All the oddities of relativity are simply consequences of this advance of realising the constancy of the speed of light to all observers.

It was realised by Einstein that the above considerations mean that one cannot simply add speeds on in an obvious way. We mentioned the story of the passenger on a plane walking while holding a torch shining ahead, it would seem obvious before relativity that one should add the walking speed to the speed of the light to find its new speed. Normally-in Newtonian theory we just add on the speed of the frame we use (the ship, the train, the plane). How can the speed of light be identical to all observers? To take an example, if a car going at 50 km/h passes a bicycle going at 20 km/h in a parallel direction, the speed of the car seen by the cyclist is 30 km/h. Most people will conclude without thinking much about what we are assuming when we add the numbers, so why does this argument not hold in the case of light beams too? Because that now contradicts our new understanding that the speed of light is fixed. The solution to such puzzles demands something radically new. If we accept Einstein's postulate, that the speed of light is fixed to all observers in the manner that science forces us to, this means that something has gone badly wrong with our intuitive ideas of adding speeds. But where could it have gone wrong? Speed is distance divided by time. Well, distance is just distance—once we decide on a standard ruler each person can check our rulers directly against the standard distance that we choose and then there is little scope in any argument. Einstein realised that the issue centred in our intuitive or classical notion of time. Time is a measure of progress with which every one has to agree but a measure that depends on how we go about measuring it, even on such matters like how we go about synchronizing our clocks. This raises the key matter, How do we know what the time is somewhere else? Every one in physics from Newton and before had assumed there was something universal about time, like space, both existed in their own right absolutely. Einstein is challenging these assumptions. Synchronising of clocks demands a properly defined scientific procedure, one way is to set up set up a standard clock in London say then freight it or a copy anywhere else where we might want a standard, and compare the local clocks and the master clock. That might seem a good an obvious way of solving all time problems but even in this way, there are many questions that need to be asked before we can be sure the result is a good one. How do we know that the freighting process did not affecting the time reading? There is a long and technical story behind this that I am not bothering the reader with in his chapter, an outline is given in chapter 4, but the upshot is that even a 'slowly' moving clock suffers significant time dilation, which affects one's choice of synchronization. So this method of slow clock transport is not as natural and straightforward as first appears. The following method is more fundamental. In practise nowadays we all use our wrist watches. A simpler method to check them all against one another is to listen to the pips from the master clocks on the radio, these arrive at the speed of light, and light obeys relativity. So we can assume Einstein's postulates that the speed of light is fixed and does not depend on the speed of the source then we can use that speed plus a measurement of the distance the light has travelled to allow for the time it took to travel, one can check on that by sending a light pulse back again. So now we know, given one standard clock, and can find by experiment what the time is everywhere else.

Time Dilation

Once the process, of setting up time standards is understood we can start to examine the more subtle things of relativity. One is time dilation, which says quite simply that moving clocks run at a slower rate than stationary clocks, if you allow for the novel ideas on the constancy of the speed of light (c). To see this, first suppose that a simple clock is made from light bouncing between two mirrors separated by a distance l (next figure).

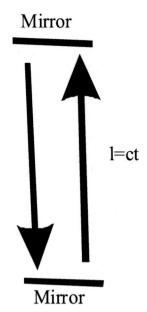

Figure 32, A Light clock

If we just call every double bounce a tick of the clock, the distance covered is $2l = 2ct$, c the speed of light, t the total time a tick takes.

This means that a double bounce will take a time 2/c. Suppose next that one were to move the clock sideways, gently at a fixed speed v; now a tick of the clock looks a bit like Figure .

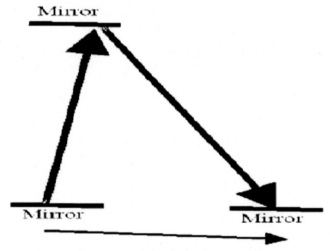

Figure 33, A light clock when moved

Now, the light will appear to have covered a longer distance than $2l$ on each bounce. But if we believe with Einstein that the speed of light remains still c, then it must have taken a longer time than $2l/c$. Pythagoras then tells us that the light is actually moving an extra distance so that one must calculate the extra time it took at speed c of course. The conclusion becomes that a moving clock seems to run more slowly than a stationary clock by a factor $\gamma = \frac{1}{\sqrt{1-(v/c)^2}}$ that comes from the Pythagorean formula to get the extra length involved. This gives the idea of time dilation. It was verified in very accurate experiments, for example,[3] which quote an accuracy of 4×10^{-5} confirming the prediction of relativistic. This is a radical Proposition, one must think of applying it to the slowing down of any clock, including our biological clocks. If that were not so, one could violate the principle of relativity. Even a biological clock may be checked for constancy and synchronisation against any agreed master clock in principle by the methods mentioned above of using the radiopips, and any discrepancy would have become obvious by now. I believe that as a physical scientist because we are made up of atoms, and I know that atomic structure is by now very well understood and is subject to the laws of physics, including relativity and quantum mechanics and must obey those laws as must the workings of any other clock. To that extent I am confident to speak on these biological matters. I will give you another reason for my confidence that goes to the roots of what we have learned about relativity so far. If some types of clocks were exempt from time dilation, say, because they were biological, then this would lead to a readily detectable violation of relativity. An airplane traveller having a biological clock in the aircraft, his heart beat, all his bio rhythms, and then compare them with his wrist watch, and perhaps trying to making allowance for effects of time dilation, could learn from that how fast the aeroplane was going in some absolute sense. The more this is considered, the more puzzling it seems to get!

Twin paradox

One peculiar consequence of time dilation is if we suppose we have two twins, Andy and Bev. Andy stays on earth, Bev goes to α-Centauri (which is the nearest star to the solar system, one of the pointers to the southern cross) by a spaceship. When she gets there, she reverses her rocket, and returns home. Then, Andy sees Bev's clock apparently running slow in both directions because of time dilation. The turnaround takes negligible time in comparison. Andy now thinks she comes home younger than he is. She has aged less, having had fewer heart beats, by *his* reckoning, than he has on *hers* since they last met. The twins (born together) now have different ages! True, no one has done that experiment with live people, since travel to the stars light years away from us is not practical. Obviously, such things were intensely discussed in the early days of relativity, and are now part of orthodox science, relevant experiments having been done with particles as mentioned later.

Mass changes

Similar and related discussions lead to the idea that the mass of a moving object looks larger than that of a stationary object, again by a factor . This means that the effect gets bigger as speed goes up, to the point that finally at the speed of light ($v = c$) the mass of any particle becomes infinite. When energy is added to a body by increasing its speed; its mass changes accordingly. It is possible to derive all this from the notion that time dilation operates in particle collisions, that mass is conserved and that force is mass times acceleration. I am aware that there is a fashion in physics teaching today to by-pass the concept that mass changes with speed. As far as I understand it, such a choice is not obligatory in teaching, and it is not always helpful in my opinion. This will be illustrated in the topic of hidden momentum in chapter 10

photon mass

a photon of frequency ω has an Energy $\hbar\omega$ and so a mass $\hbar\omega c^2$ given the standard conversion $E=mc^2$ between energy and mass in special relativity, this squares with the results of many thought experiments in electromagnetism. Even if a particle is at rest, it has a minimum rest mass m_0 and so stored energy. This rest mass energy can supply the energy of radioactive decay, and so the associated nuclear energy. (Some unfortunate Anti-Einstein publicity in 2011 on such matters is not yet independently verified. and in fact the claim has been retracted, having been traced to defects in the associated experiments)

How do we know that such bizarre predictions of relativity are in fact true?

These various items have been tested so often and so accurately that no doubt can be held now of them. For example, chronometric accuracy has improved by nine powers of 10 since 200 years ago when Harrison won the prize for constructing a chronometer accurate to 1 sec per day[4]. Today we have mobile clocks accurate to 1 microsecond per year. They are set into clock synchronisation worldwide essentially by Einstein's radar 'light' flash method, with relativity corrections applied for the time of light speed travel. If they were wrong after all, the world's shipping would be off course by several kilometres—enough to give several more shipping accidents per year[5]. In 1971, Hafele and <u>Keating</u>[6] flew atomic clocks around the world in standard commercial flights. They verified the twin paradox: the moving clocks aged less than the ground-based clocks that had been set up identically and synchronised, before the flights there. In effect the clocks were born twins. They also verified other relativistic effects of time dilation explained above, though they also were affected by the general relativistic effect of the red shift arising from light moving in a gravitational field. Magnetism can be viewed to be a peculiarly relativistic effect. This goes right back to the symmetry Maxwell discovered between electric and magnetic effects. If instead of saying Electricity + Magnetism $\Rightarrow$ Relativistic Optics, we turn the logical equation around and make the logical sequence to read Electricity + Relativity $\Rightarrow$ Magnetism. I mean by this that on moving electric charges, their speed automatically gives rise to magnetic effects, if one makes proper allowance for length contraction, time dilation and so on arising from the speed change. This tells us that we should not dismiss relativity as merely a small thing, just a change in decimal places on a sensitive clock. The Magnetism force exists at an obvious tangible classical level, compass Needles turn, nails stick to bar magnets and so on. All this is despite the fact that drift velocities of electrons, in a current-carrying wire are very modest by relativistic standards. The full answer here is very complicated and involved the magnetism locked inside the particles making up the atom, in fact magnetism is an intrinsically quantum mechanical effect as well as a relativistic effect, and this whole matter is obviously too long for us to explore. It might seem preposterous to suggest that we need a relativistic effect. This also means that electric motors and generators that also depend on magnets all work only because of relativity. Relativistic corrections are in fact of direct use and importance in such things as designing klystrons (microwave generators) and Electron microscopes. Where it is essential to include relativity corrections for successful design, modern particle accelerators prove relativity every time they operate. An electron can today be accelerated to energies of the order of giga electron volts (10^9 eV) compared with its rest mass energy of 0.5 MeV (5×10^6 eV). Given that 99.95% of its energy is kinetic energy, 0.05% rest mass energy, it is tramping at very close to the speed of light: ~99.999999995% of c in fact. Again, the design of these devices would be unsuccessful if relativity were not correct. This all means that now the truth of relativity depends not on Einstein's theory of gravity, which replaces Newton's theory, it also has its own portfolio of convincing experimental vindications. In Einstein's General theory of relativity a new theory of gravity is fully included. One General Relativistic effect will concern us in this book for ring laser physics that will be treated here, if if a gravitating body is rotated at a rate Ω, the local inertial frames nearby are also rotated at the rate h=rsΩ cosλ/5R

where rs is the Schwarzschild radius $2GM//c2$ and R t he radius and λ the latitude this modifying the Sagnac effect, that is it changes the beat frequency between counter circulating waves.

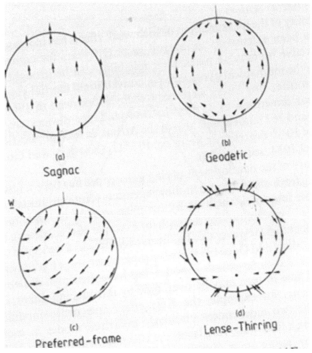

Figure 34, Contributions to the Sagnac signal

The diagram above indicates the dependence on latitude of the gyromagnetic fields associated with the (a) the sagnac, (b) geodetic (c) preferred frame (c) Lense-Thirring frame dragging respectfully[7]The feasibility of detecting the Lense-Thirring effect with modern ring lasers was first discussed by Stedman, Schreiber, and Bilger in[8]. Superbly engineered mechanical gyroscopes went into orbit in October 2004 to test this prediction of Einstein's General Relativity in space, the test was successful.

Practical ways of solving the problems of detecting the Lens-Thirring effect from earth rotation in, a ring laser very well defined geometry have since been offered by (Pisa group)[9] and by Hurst[10] see Chapter 7 for an outline discussion.

At several times including recently false assertions have been made about the Lense-Thirring effect type frame dragging in relation to rotating superconductors and one such assertion that has concerned us recently is that Tajmar hypothesized[11] frame dragging in rotating superconductors, generate a Gravito-magnetic flow detectable for example by a ring laser as a component of a Sagnac signal .This is not the first bizarre and unfounded prediction made about the behavior of superconductors in gravitational fields some risks in this are described in .[12] There has even been talk for example of superconductors shielding gravity[13] And it was suggested that oxide superconductors can levitate under rotation[i] These examples should suffice to warn the serious physicist from rushing into such ideas. The fact is that our understanding of superconductors since the BCS theory makes them no longer the exceptional mysteries some paint them to be. this aberrant theory of Tamjars was in practice a simple casualty of a straightforward search with the Canterbury ring laser. For their satisfaction, and in the spirit of science a local team did a negative search in the Cashmere Cavern showing no such effect existed as had been proposed this could fairly recognized as one clear application of the ring laser project to a project supposedly involving fundamental physics albeit of a very dubious character.[14] The concept of rotating heavy masses near optical interferometers is at least as old as Lodge (see the preface) and has been seriously considered by others [15]

The timing system we use at Cashmere, like all uses of GPS, absolutely depends on the corrections to Newton afforded by Einstein's theory, and would fail within minutes if relativity were untrue.

The de Broglie waves of atom gyroscopes etc. were predicted by the application of elementary relativity to within quantum theory.

Only relativity gives a proper answer to questions such as: What defines the "non-rotating frame" in which a gyroscope rotation sensor detects zero?

The solar system orbits the centre of the Milky Way galaxy at a linear speed of 250 km/s, The milky way rotates at 1×10^{-6} °/an (/an)= per annum the universe rotates 1×10^{-10} °/an. Light is such an extraordinary and wonderful phenomenon that a ring laser sees through all of these rotations to a frame at absolute rest from rotation. What it generates is then a measure of absolute rotation.

Endnotes

[1]   Harvey R. Brown Physical Relativity Clarendon Press Oxford 2005.
[2]   Brillet and Hall (Phys. Rev. Lett. 42 549, 1979)
[3]   Kaivola et al, Phys. Rev. Lett. 54, 255 (1985).
[4]   Sobel Dava Longitude The True Story of a Lone Genius Who Solved the Greatest Scientific Problem of His Time Fourth Estate Ltd, 1998.

[5]   (See Cohen et al, Phys. Rev Lett. 51 1501 (1983), 53 1858 (1984).)

[6]   Science 177 166 (1972).
[7]   GEN F.
[8]   Stedman Schreiber and Bilger Classical and Quantum gravitation 20 2527-2540 (2003)
[9]   F. Bosii et al. Physical Review D 84 122002 (2011)
[10]   Marsden grant proposal 2011.
[11]   arXiv:gr-qc/0607086
[12]   Payne and Stedman Physics Letters 50B 415-416 (1974)
[13]   G Hathaway B Cleveland Y Bao Physica C 385 2003, Pages 488–500 (1974)
[14]   R.D. Graham, R.B Hurst, R.J Thirkettle, C.H Rowe, P.H Butler. Experiment to Detect Frame Dragging in a Lead Superconductor. In Physica C: Superconductivity and its applications, **468** 383-387.
[15]   . G. E. Stedman K.U. Schreiber and H. R. Bilger the detectability of the Lense-Thirring field from rotating laboratory masses using ring laser gyroscope interferometers}, Classical & Quantum Gravity 20 2527-2540 (2003).

6 Quantum Mechanics

INTRODUCTION

If relativity is peculiar from a classical viewpoint, and if its predictions are highly counter-intuitive and mind bending, Quantum Mechanics is, if anything, even more so. Both these revolutions, dating from the 1920's show that our classical prejudices regarding time, space and matter are wrong at a fundamental level. Murray Gell- Mann (a Nobel Laureate in physics was a key discoverer of the quarks structure of the proton and neutron when at Canterbury University 1978) said, "Deep down inside, no-one understands Quantum Mechanics." If one thinks, "why bother with Quantum theory then?" you might say the answer is. "Because that is what Nature requires." Another Nobel Laureate indeed classmate and colleague of Gell-Mann, Richard Feynman, once said in Robb lectures at Auckland University (New Zealand, 1979)[1] "I'm going to describe how Nature is,' 'and if by any chance you don't like how nature is, that's going to get in the way of your understanding it.' 'If you don't like it then you can go live somewhere else.' [i.e. some other universe] 'I can't explain why Nature works in this peculiar way. Again we are forced to face with the reality of the world around we have no choice but to accept it'. These are problems that physicists have learned to deal with and live with of necessity of course, for we live here. If anyone puts the question of what the essential content of Quantum Mechanics actually is, it centres on a set of elaborate mathematical recipes. Armed with these, one can in principle compute the result of any experiment with small particles. The first approach of a physicist is to learn the vital mathematical rules, then verify them in checking experiments and so learn to use these rules. And indeed we physicists believe that we have found them and that they are unique, and adequate for all experiments that one can do today.. In the work of Feynman and his book 'QED'[2] he was actually at some considerable pains to help a new physics student from becoming too involved with some of the deepest questions in the topic. Feynman described in outline the full machinery of quantum theory using an extraordinary and simple pictorial construction. He presents various geometric contractions involving the adding of a lot of arrows head to tail to show why light moves linearly, reflects at a mirror and does all the other peculiar wavy things that a light wave does. The construction has the effect of avoiding even the idea that the square root of −1 is a number an idea essential for quantum theory. When Feynman explains this he said "No matter how many arrows (amplitudes) we draw, add or multiply, our objective is to calculate a single final arrow, for the whole history." I will not attempt to explain this, (see the earlier ref to Feynmans book QED) that would have to assume far too much mathematical ability and physical science knowledge for the general reader. Feynman's account is one place to go for as simple an exposition as possible. The story of Quantum Electrodynamics is absolutely fascinating, with a majestic agreement between accurate experiment and detailed theory, making quantum theory the most successful theory of nature ever: again Feynman's book[3] sets these advances in context. The heyday of such work has been post World War II, and the full development of the field theories of Quantum Mechanics has extended to the 1980's. The majority of applications of Quantum theory do not require a full discussion of these more recent developments. The rules of the theory inform the physicist how to conduct and interpret any experiment, and how to formulate the necessary mathematical analysis. What I am offering in this chapter is a simple outline and introduction to a few results germane to quantum theory.

Almost any book on Quantum Mechanics will begin the topic and soon start talking of the Copenhagen interpretation of Quantum Mechanics, which is from the school of physics established by the Danish physicist Niels Bohr of Copenhagen. The key issue here is that Quantum Theory has taught us that all particles have wave properties at a microscopic level, so that these wave and particle pictures are "complementary" to each other, for an understanding of the subatomic world. For centuries since the time of the ancient Greeks a debate has raged over whether a beam of light was to be thought of as a stream of particles or a vibration or wave; these were natural classical models. The wave theory appeared best with the monumental developments of the 1800's with the discovery of polarisation and then of the Maxwell theory of light (Chapter 5), both clear evidences of the wave properties of a light beam. During the early 20th century, there was strong evidence that the particle model had truth also. Finally, physicists came to recognise that neither model won outright but rather both models had essential truth, and both were needed to give the whole truth, and to explain the results of all experiments. And the problem facing Bohr Einstein and others within the early days of quantum theory was to find out how to make these two aspects of truths compatible with out sacrificing one or the other. Not every one now agrees with Bohr's particular understanding of the resolution of the paradoxes that Quantum Mechanics brings, and one prominent person who disagreed was Einstein. It is a surprise perhaps that Einstein was so opposed to a theory to which he laid the foundations, when he discussed the photon or particle theory of light. There were many simmering debates within science over the nature and fundamental interpretation of the mathematical rules of Quantum Theory. The 'Copenhagen' interpretations of Bohr *et al.* held sway for decades. Others prefer to follow a more overtly pragmatically approach, and follow Richard Feynman and Murray Gell- Mann, Nobel laureates from Caltech and outstanding pioneers for some of the most esoteric areas of particle physics, who yet tended to eschew any deep philosophies on why the rules work. I call that the Caltech interpretation of Quantum Mechanics. The possible mathematical recipes are not nowadays disputed, and all have thoroughly tested for that; let me put the question as to how to understand the absorption of light particles.

ATOMIC SPECTRA

In the very early days of quantum theory, the problem was to understand why gases of hot atoms had distinctive colours, and very sharp spectra (colours), the wavelengths being very precisely defined, so that such gases absorbed and emitted just certain colours of light and no others. Bohr stumbled on the right kind of approach, the whole atom jumped from one state to another when emitting light, with an energy corresponding to the change in the energy of the atom. Rutherford[4] showed through his α-particle experiments that the atom consisted of a central heavy nucleus with electrons orbiting it like a miniature solar system. Einstein identified the light emitted from the jump to be a particle, of a definite energy and mass. Bohr then postulated that the colours could be explained by the occurrence of sudden changes in the electronic states of the atom as those for which one whole 'quantum' or 'particle of light' was swallowed or emitted. The difference between the electron energies between these two states was precisely defined and explained, by energy conservation, through the precise colour of the light involved. Although Bohr's original theory of this has had to be redefined a lot, the basic idea is now a cornerstone of all modern science.

LIGHT

As for relativity, the behaviour of light is of central importance within Quantum Theory. Some like Newton already thought that light was a stream of particle. Other realised it was also a wave Eventually with the advent of the Quantum Theory of light it was realised that both models had truth light did involve a flow of particles. At yet at the same time light must also be understood to be a wave motion; so that a light beam does have a frequency f and corresponding to it a wavelength λ, the distance between wave crests related to f by $f\lambda = c$.

Either of the wave param-eters ω, λ serves to define the colour of a light beam and the energy of a light particle being $E = \hbar \omega$ or $\hbar c / \lambda$. $\hbar$ is an incredibly small constant of nature $\hbar = 6.6260755 \times 10^{-34}$ (Joule seconds). One odd irony within this story is that Newton 1643–1727 himself had good evidence in his observation of ring when a watch glass is placed on a flat surface that light was a wave, but one of the most convincing proofs that light is a wave is that it can be polarised. Polaroid sunglasses and the like allow restriction of the wave motion of the electric field of a light wave to just a single plane, the plane of polarisation. Polaroid was the invention of a first year student, Edwin Land, who never got a degree. Remember that Maxwell's theory of electromagnetic waves allows the whole range of electromagnetic waves which include X-rays, microwaves, radio, ultraviolet, γ etc.,

gamma rays
X-rays
Ultra violet

infra red
microwave
television
radio   101

Figure 35, The electromagnetic spectrum

is based on the ideas that moving electric fields, which create magnetic fields and vice versa, a combination of two such actions result in a wave travelling at the fixed speed $c = 299,792,458$ m/s.

PHOTONS

For all that Maxwell found the reason for the wave nature of light, Feynman emphasises for his masterly account in QED that light is a particle stream of photons. We need to spell out a little more. Once the Rutherford atom was understood (1911 was the discovery of the nucleus), it was soon realised that the problem of atomic spectra could be understood. In that the energy of the light particle released as the atom jumped between different electron states. This energy corresponds to the energy difference between two electronic states associated with the jump that released the light particle, of corresponding energy Within relativity you can find the mass of the associated particle of light by using the relativistic equation $E = m_0 c^2$ by putting these last two equations together we get the mass of a light particle $\omega > f$ viz $m_0 = \hbar f / c^2$. It turns out of course that light particles are very small. This can be seen from the last equation once you remember that $\hbar$ is a small number and c is very big. The idea of such a size to a small packet of light is what gave rise to the name Quantum and it characterises the whole theory. All this means is that light might best be thought of as being a pattern of vibrations of electric and magnetic fields through all space (electric and magnetic fields all travelling through space at the speed c). For some experiments a light beam must also be thought of as a stream of particles flowing. Each of the wave and particle picture is equally and importantly true, since each is confirmed by experiment.

63

It follows that physicists found themselves forced to try to reconcile these two different pictures—wave and particle pictures of light. This led to all types of quite diabolic subtleties of interpretation, with some of which we shall have to wrestle. For practical situations, both pictures are vital to understand experiments. A 1 mW laser beam of green light carries 2×10^{15} photons per second. A 100 W light bulb produces approximately 10 billion-billion per second. Various estimates have been made of the number of photons it takes for the brain to register it. The number is amazingly few. One has to remember that 10% of the light entering the eye is reflected at the cornea, and 47% absorbed inside the eye. It has nevertheless been claimed on the basis of evidence that the human eye can detect just one photon. It is an astonishing thing when you remember that the same detector, our eye, can cope with the enormous extremes of sunlight and starlight down to one photon[6].

POLARIZATION

Waves such as light waves can have the property of polarization, this has to do with the direction of the wave motion. Waves on a string can pass through a slit through which the string itself passes if the direction of the slit matches the direction of the motion of the string. This is just as true for light waves as for string waves. The slit then

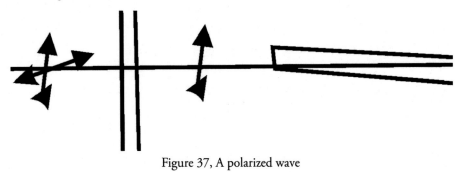

Figure 37, A polarized wave

Figure 36, An electromagnetic wave

corresponds to a piece of polarizing material. A light wave that is polarized has its electric field in a particular plane. This plane is specified by the laws of reflection. The polarization of light, was recognized during the early 1800's from such properties. It was these developments that lead to the triumph of the wave theory of light over Newton's old corpuscular theory during the 1800's. It is one of the ironies of science that the Quantum theory has forced us to return to the particle model of light to explain some experiments. Now let us consider in more detail events going on when a polarized light beam hits a Polaroid sunglass. The events occurring at the particle level depends critically on the angle θ the particle electric fields make with the axis of the Polaroid, that is to say the direction in which the electric field must have if the light is to be transmitted, we shall make that the direction of the slit in the following figures. In Figure 38, the sequence of events we consider is that a light wave from the left encounters a Polaroid at (a) that makes it polarised (b), it then encounters another Polaroid, tilted at an angle θ it emerges if at all with a polarisation matching that of the second Polaroid (d) but with an intensity reduced by $\cos^2\theta$, This is called Mallus's law. When a particle of light goes through two Polaroid's tilted at angle θ both Quantum Mechanics and classical wave theory predict that the chance that it will go through both is $\cos^2\theta$. In a particle goes through a slit. It is most likely the particle will come out immediately opposite the slit. This is because the wavelets interfere with each other reducing the chance of detecting a particle, because

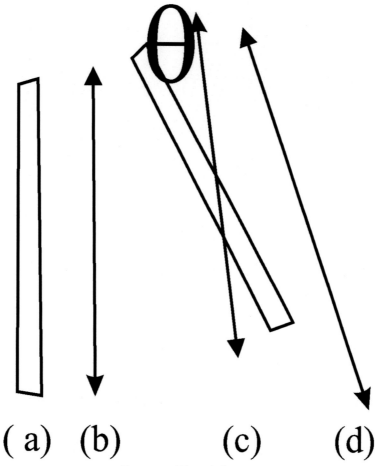

Figure 38, Theta definition

the various waves from the slit are interfering destructively at these positions. A 'mobile' graphical depiction of this showing the colour dependence is available at various web sites.[7][8][9] In the case of a double slit or a screen with two holes the pattern becomes correspondingly more complicated again.[10][11]

Once again, there are certain angles or directions on the right side where one cannot easily detect a photon because the associated waves from the two slits interfere destructively. One again, mobiles are available so that one can see the effect of changing colour, slit size or screen spacing.[12] Then, one is faced with the baffling and deep question, if you detect a particle of light, which slit did each particle of light go through? The obvious sensible answer would be that the particle only went through one slit. Yet the wave interference pattern tells us that it or at least the wave associated with the particle went through both. The detection statistics prove that. But how can one particle go through both of two holes? The wave associated with it goes through both holes and helps guide the particle to its destination. Photons are never found when a light detector is placed at the dark parts of the above diagram. Apparently each particle of light knows that both holes are there. We have learned that it is pointless to try to solve this mystery, somehow each light particle knows both slits are there, and somehow something linked to the particle went through both. On the right side of the holes sufficient information was available to build up the whole interference pattern. One can imagine clever ways that one might try to find out the hole the particle went through, but it turns out that nature conspires against us, and that the very attempt will kill the interference pattern. Any attempt at finding out which hole by extending the experiment in some way will have the effect of killing the whole wave diffraction effect. Some ingenious thought experiments, of this kind have been proposed by Otto Frisch[13] for example. All turn out to have unavoidable side effects. These contrive to kill precisely the very effect one is trying to measure. There seems to be a

Figure 39, John Bell

principle in the universe to obstruct our desire to probe it fully. One reaches the somewhat infuriating conclusion that any attempt to try to "understand" any quantum mechanical experiment in detail, is futile. What this tells us is that classical modes of thinking, based on ideas like a particle can only go through only one hole within a screen, for all their common sense appeal, show that our classical ideas simply fail for tiny things a fascinating article by Frisch explores the idea of generating a single photon and checking it[14].

In relatively recent times it was found that strong theorems exist disallowing for ever learning all the information that classically one would like to obtain from quantum theory. John Bell discovered[15] one key such result in 1964 which I illustrate using a lecture demonstration apparatus. depicted below is fully explain in a book defending science[16] The primary source in a journal article[17]

MASS WAVES

Another of the amazing results of Quantum Mechanics is that *all* particles, such as electrons, as well as the neutrons and protons that make up atomic nuclei, diffract like waves and have a behaviour governed by wave rules within Quantum Mechanics. The wavelength of the wave associated with any particle is related simply to its speed v and mass m by the law since

$$\lambda = \hbar/mv. \quad (1)$$

This law applies in principle also to footballs, bricks, but the wavelengths involved are incredibly small since Planck's constant is so small. This means their quantum properties are negligible and football fans need not worry concerning quantum diffraction effects of the football, for that will be no reason why your favourite team lost their last match. Electrons are readily diffracted and such quantum effects are the reason one can make electron diffraction microscopes, to take one example. Again it might seem obvious that an electron passing through a screen with two holes must have gone through just one of two holes before it lands on a fluorescent screen, but one can not specify the particular hole it went through nor where it will land on the screen, the landing point depending simply on the laws of chance that are determined by the strength of the electron wave at each possible location on that screen. If the electrons pass through a single slit to arrive at the screen, the pattern of points at which a flood of electrons arrive builds up the characteristic diffraction pattern of single slit. The picture above shows the 'barrage pattern' building up over time as the electrons arrive from the slit at a detection screen after going through a simple slit notice the vertical black fringes in (see the interferogram on electron diffraction in Wikipedia.) [18] The white bars within the diagram indicate the angles in the diffraction pattern where no particles are ever seen because the associated waves interfere. These correspond exactly to the forbidden angles for the photon case, when allowance is made for the change of wavelength. The full theory of Quantum Mechanics gives precise algebraic functions for the probability of a electron or photon arriving at given time at a given point in such diffraction patterns. That is where the full formalism of quantum theory cannot be avoided. The same phenomenon has been well studied in neutrons [19], and even whole atoms

Endnotes

[1]   In his Robb lectures in Auckland NZ, DVD's can be purchased, http://www.vega.org.uk/video/subseries/

[2]   the ensuing book. Richard P. Feynman. QED: The Strange Theory of Light and Matter (Princeton Science Library) Penguin Books Ltd 1990.

[3]   the ensuing book. Richard P. Feynman. QED: The Strange Theory of Light and Matter (Princeton Science Library) Penguin Books Ltd 1990.

[4]   50 a recent biography is Rutherford Scientist Supreme, John Campbell AAS Publications, 1999

[5]   Wikipedia electromagnetic spectrum.

[6]   F. Bieke and D. A. Baylor Rev. Mod. Phys.**70** 1027-1036 (1998).

[7]   http://www.physics.uoguelph.ca/applets/Intro_physics/kisalev/java/slitdiffr/index.html

[8]   http://www.physics.uoguelph.ca/applets/Intro_physics/kisalev/java/slitdiffr/index.html

[9]   http://www.colorado.edu/physics/2000/schroedinger/two-slit2.html

[10]   Double-slit experiment Wikipedia.

[11]   Wikipedia Double-slit Experiment

[12]   http://vsg.quasihome.com/interfer.htm

[13]   O R Frisch Contemporary Physics **7** 45 (1965)

[14]   O. R. Frisch, Contemporary. Phys. 7 p. 45 (1965).

[15]   For one introduction see Physics world June 2000 p7 J S Bell Physica 1 192 (1964)

[16]   I have described my own lecture demonstration of the issues at stake using the language of polarised light for an elementary understanding of the demonstration in 'An orthodox Understanding of the Bible with Physical Science G E Stedman Strategic book Co. 2012 appendices.

[17]   The Primary source in G E Stedman American Journal of Physics **53** pp11414 (1985)

[18]   Particle diffraction Wikipedia

[19]   Neutron Interferometry Lessons in experimental Quantum mechanics H. Rauch and S Werner (Clarendon Press Oxford 2000)

7 The Farm of ring lasers

General introduction

two quick comparisons

First I give one quick comparison a Musical scale for the earth rotation frequency generated by some of the lasers considered in this chapter which all gave Sagnac signals (chapter 2) in the audio frequency range.

Laser	Sagnac frequency (Hz)	Approx. on Equi-tempered scale
C-I	69.3	$C^{\#}_2$
C-II	79	$D^{\#}_2$
G0	288.3	$D^{\#}_4$
G	348.5	F_4
UG1	1512.8	$D^{\#}_7$
UG2	2177.6	$C^{\#}_7$

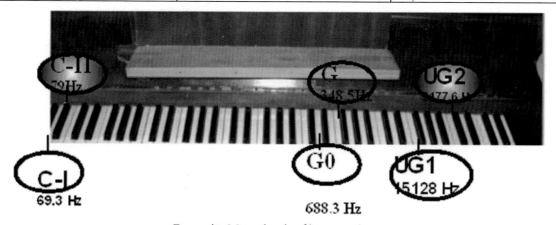

Figure 40, Musical scale of large ring lasers.

The frequencies listed pertain to the Sagnac frequencies induced by earth rotation on a ring laser interferometer, This musical link is not just a joke. Clive Rowe brought an old amplifier unit to the Cashmere cave to play frequency generated by earth rotation in each device because harmonic distortion and main hum are more sensitively detected by ear. Clive comments. "I found the sound helpful when adjusting power, mirrors and other ring-laser parameters as the background sound gave useful information on the Sagnac signal and harmonics. It also gave a clear indication of mode jumps and mechanical disturbances. The sound output of the FSR beat frequency also assisted in adjustment when eyes were looking at other data."

In such lasers, the light beam bounces along a polygonal line between mirrors. (We construct the Sagnac interference pattern from the light that leaks out the mirrors at each reflection) Both clockwise and counter-clockwise beams lase. As a result of earth rotation, the earth motion moved during the very small transit time of the light beam in fact. Rotation makes the co-rotating beam slightly redder and the counter-rotating beam very slightly bluer. A ring laser can measure the absolute rotation of whatever environment it is placed in.

With better mirrors and bigger dimensions, the device becomes increasingly sensitive to rotation (Chapter 2)– but also correspondingly much harder to construct and operate.

Another quick comparison[1] of particular importance for characterising the excellence of frequency sources is in the Allan variances, a severe and stringent indication of their fundamental stability limitations as clocks. Given the existence of all kinds of noise sources degrading their performance.

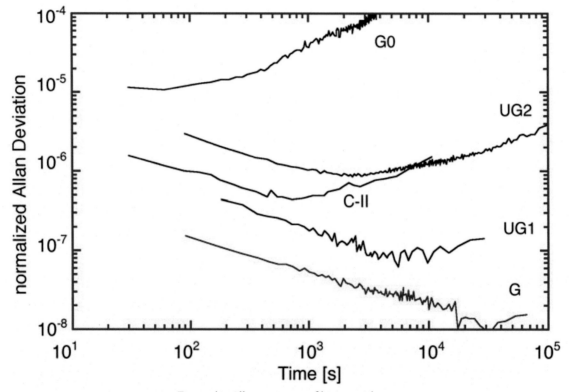

Figure 41. Allan variances of large ring lasers

A comparison of the resolving powers of the Large ring Lasers to be introduced in this Chapter is taken from their observed experimental behaviour. These are plots of the Allan deviations for C-II, G0, G, UG1 and UG2 normalized by their Sagnac frequencies. For all the rings we plot Allan deviation normalized by their Sagnac frequencies. The low-time asymptotes indicate quantum noise limitations, higher times show the effect of many noise sources.[2]

The farm, introduction

The Canterbury project initially built a simply constructed ring C-I. This machine achieved a sensitivity down to parts per million of the earth rotation. A German group then joined us following the C-I success. Building a better engineered laser in Germany for installation at Cashmere Christchurch during 1997 this was also a one-metre square ring laser, dubbed C-II (for Canterbury II), this machine had a monolithic construction in a half-tonne slab of low-expansion ceramic glass (Zerodur). a substantial budget was required. The aim was to sense rotational motion down to a rate of parts per million of the earth rotation rate a performance which was routinely achieved.

The ultimate aim for the German group was a machine of their own i.e. for installing in Germany called G or Gross-ring 4 metres square that could achieve a few parts per billion of the Earth rotation rate. The success of C-II and of other Cashmere laser G0 led to approval for proceeding with G. G0 was another proof of principle for G. A pilot ring, of similar area to that planned G, G0's area was in fact approximately 12 square metres it was mounted vertically the only available table at Cashmere, placed a concrete wall left by the military. Like C-II G0 was placed at Cashmere in 1997. G0 machine can reached 10 parts per million precision for several hours, despite its outrageously cheap construction.

Again G would be built from a solid block of Zerodur, from Schott Mainz now of diameter 4.2 m, as its major mechanical component. The bunker for G was built specially at the research station at Wettzell, Bavaria, Germany, the Zerodur slab being supported on a concrete pier itself going down several metres into bedrock and with careful protection against temperature and air pressure drifts. The decision to proceed with this venture followed the success of the C-II and G0 projects described here. After these the Canterbury group with assistance from collaborators embarked on still larger machines UG-1 UG-2 simply placed on bedrock whose areas were substantial fractions of the total cave area at Cashmere.

Risks of success in these projects

At every stage of these developments the projects involved substantial risks as regards scientific success. As mentioned in chapter 2. At the C-I level one risks included a popular perception that it exceeded the perceived practical limit to feasible ring areas which. All of the ring laser devices were limited in practice by quantum shot noise to areas of typical for navigation gyros viz 60 cm^2 in fact due to the difficulty in maintain good mirror cleanliness shot noise limits were very far from being a problem. In C-I or indeed all our large lasers.

A more serious risk was the need to maintain single mode operations, several cavity modes would be within the gain curve of the light amplifier. Our trick was simply to reduce the power levels until neighbouring modes by reducing the gain so that unwanted modes fell below the gain curve and were simply starved out. One common aid is to use isotopically enriched neon natural Ne in 90% Ne^{20} admixing Ne^{22} changes the gain curve helpfully in ring lasers, we did this as soon as we could afford it. Another risk with C-I was that backscatter etc. might be so strong as to make earth rotation alone in sufficient to unlock the ring this proved not to be a significant problem.

In later machines we also used Bob Dunn's trick for mode control of exploiting the effects of changing recalled the laser line shape by increasing the gas pressure so that the Doppler broadening of the laser line so that over-pressuring the He-Ne lasing gas gave an a very effective control mechanism especially in the larger lasers, all these lasers would of course also have many longitudinal cavity modes were even more densely packed and mode discrimination correspondingly even more important however. Our risks in going for the UG series of lasers would have to be rated as 100% i.e. near certainty fatal given their simple construction and lack of any elements whatever stabilising them against temperature shift (some trials were made with well compensated mirror holder's). I thought it a gamble worth tacking for what we might learn, about the possibility of mode control using the above methods and, given above every for such gargantuan rings. In the event backscatter effects made us shut UG2 down a year or so after construction.

C-II and G0 and UG1measured effects never measured before by such techniques. One is the rotation accompanying the mantle waves from earthquakes, even when the epicentre is on the other side of the world. Large events trigger mantle waves of period 22 s, We also observed and explored the tidal effect of the Moon on the solid Earth. So Seismic studies at Cashmere have proved to be very novel and exciting. Virtually no data was available previously to either seismologists or engineers on the important rotational effects accompanying earthquakes. C-I measured rotational effects from several sizeable regional earthquakes.

For the first time tele-seismic wave rotations have been detected in 1999 at Cashmere using C-II and also G0. The expected enhancement in sensitivity with the UG series of lasers is more than proportional to the area increase, allowing fuller study of seismic waves and detection of the many more earthquakes whose magnitudes are much lower on the Richter scale than those studied so far. The 3rd international workshop on seismic rotations 3IWGRS was held at Christchurch in September 2013. The presentations are available on line and are now also published by Springer [3].

I do not attempt to cover in this book all related history there are few other groups world wide which are active in this kind of research. Such efforts are rare, and it certainly seemed to us that we had the field pretty much to ourselves. One worthy of special mention however is that of Professor Bob Dunn at Conway, Arkansas U.S.A. recently showed that a triangular HeNe ring laser with perimeter 40 metres can operate as a gyroscope. We warmly welcomed Bob to New Zealand where with the German group we built a series of larger very simple ring lasers at Cashmere, New Zealand, the first dubbed ultra-G or UG1, which used not a Zerodur table but corner boxes from stainless steel set on concrete plinths into the ground basaltic rock. Corner boxes were connected by stainless steel pipes carrying the lasing gas mixture up to an area for UG2 of 833 m^2

As expected this machines was very sensitive, and an excellent detector for seismic rotations. In fact it discovered certain daily oscillations of the earth's axis that story we shall explain briefly later in this chapter. Following this we built an ever larger devise UG-2 with area 834 m^2 the full area of the cavern [4]. Clearly in spite of the attendant problems there appears to be no fundamental limit to the area and associated resolving power of such devices.

A ring laser senses any non-reciprocity in the effective path length of the counter-rotating beams, induced for example by absolute rotation. The two modes consist of one clockwise and one counter-clockwise mode which form a standing wave pattern, with (in the Canterbury ring) approximately 11 million nodes and antinodes, inside the cavity(Chapter 2). The Canterbury ring system C-I has a nearly square optical cavity with an area of 0.7547 square metres and a perimeter of 3.477 m (and so a free spectral range or mode spacing c/P (or 86.22 MHz.) Measuring this is a very accurate measure of the perimeter. With a He-Ne gas mixture as the medium, its two modes lasing at 633.0 nm. It has been possible to measured frequency splitting's with a precision down to microhertz level, which is 1 part in 5×10^{20} of the optical frequency of the laser.

Hans analysed the effects of mirror upon the stability of the laser this means the effects of all angular and positional alignments of all mirrors from these it was clear their precision of 20arc seconds or so was required in and 100 μm in position the possible use of a piezo support. For the mirror. Was heavily discussed but in the event was deferred for several years and never fully implemented. Similarly the use of a heater for bake out purposes was also deferred for some years.

The gas constitution was also discussed and a pressure of 0.9 torr said originally determined the gain tube geometry was analysed at great length with a view to possible appodisation.

The main argument which favours r.f. powering was the lack of drift from Langmuir flow inside the gain tube plasma and consequent drifts of a Sagnac frequency once again such matters were discussed in great detail. This was therefore favoured over DC excitation as commonly used in laser gyros. C-I was certainly the pioneer of this approach; as far as we know, we used a longitudinal coil surrounding one arm of the laser path with servo feedback to adjust the gain.

Hans suggested thinking of C-I as a small Ring with the possible size ultimately being several metres square; this was the first suggestion of something beyond C-I, in November 1987. Hans was proposing various rings of unusual shape like a square bowtie to allow electromagnetic fields in the beam-crossing region; we never got to the point of building such a machine Hans did eye he cross adits in the cave as possibilities for using larger area ring lasers, an idea which we later tested with the UG series of ring lasers, Hans had thought such would require a super-invar platform., in the end we simply used the cave floor geometry as our table, sinking important piers to bedrock where possible.

Especially in these early days, our rings mirrors ended up far worse than at their manufacture several technicians become good mirror cleaners using carefully divulged advice from the manufacturers. There are several reasons for this contamination may be offered. Our best gas pump. In those early days was a diffusion pumps and its oil leaked into the whole gas handling system the so the laser cavity our later turbo pumps, we therefore went though a steep learning curve in vacuum pump technology, before we achieved any measure of control of this problem,

For most of our work the frequency of either laser by itself was not frequency stabilized. in 1997 control of C-II was achieved in C-II that 474 THz corresponding to the red Helium-neon laser line. was stabilized by beating against a Winters reference laser. As a result we could get the absolute optical frequencies measured to 0.3 MHz.- less than one part in a billion. As part of this operation Canterbury had bought a high quality Fabry Perot for beating the two laser beams. The suppliers Newport refused to supply us direct and required us to pay US$4280 (for us a colossal sum) direct to their Australian agents. we paid the money but the agents then then went into liquidation before they delivered the goods. We appealed to all parties; who initially demanded a full payment again! we took our appeal redoubled to Newport for the honour of their good name, they agreed to deliver direct to us given our payment after some months; in the meantime the Registrar Allen Hayward of the University kindly helped us over this financial negotiation. This was a serendipitous purchase in that a few years later this device was taken off the market, became unprocurable (I believe some equivalents have become available in Europe.) The reader will understand that for a 'mere theorist' such adventures as those just described consisted financial and administrative crises I could well have done without in managing such a project.

But the Sagnac frequency from Earth rotation is far less stable again. For C-II we are talking now about variations in 80 Hz. at the level of two parts in 10 trillions of the laser frequency! Nor is this lost in the general line broadening effects that affect each line in isolation, for most of these cancel on comparing the lines. the posters at the end of chapter 2 illustrate the sensitivity of these devices on a frequency scale.

We can compare this against this part per billion frequency fluctuation that in the Sagnac frequency itself, our methodology was aiming ultimately for an accuracy of one billionth of this - one billionth of the Earth rotation rate. C-II attained a sensitivity of a few millionths of the earth rotation rate.

We may also compare these with the accuracy of our measurement of the perimeter. We put the ring in "multimode" - several adjacent longitudinal modes - which beat against each other to generate radio waves - 75 MHz for C-II. When this frequency is compared with a GPS-stabilized frequency counter which can measure the difference to an accuracy well below mHz, because the GPS satellites transmits the ticks of an atomic clock which is stable to parts in 10^{12}. So that means that one can measure the perimeter to an accuracy of 10 pm (pico-metres) - parts in 10000 of the wavelength of light.

for all that C-I was unsophisticated and unstable; we were a new group with very limited facilities and budget. However super-mirrors donated for this machine by Ojai Research sufficed to show the scientific potential of the methodology described above.

Figure 42, C-I

The Canterbury Ring Laser,was built in the Department of Physics principally by Morrie Poulton and Clive Rowe under the guidance of Professor Hans Bilger of Oklahoma State University, is a large laser gyroscope with an unusually high frequency resolution.

Figure 43, C-I 'by night' i.e. by scattered laser light, mainly from mirror imperfections and imperfect cleanliness. The RF excitation section is at the top a solenoid-al coil was fed from an r.f. tuned circuit.

This machine in the cave showed 'first light' on November 23   1989. This was later improved by better gas and vacuum handling from a diffusion pump so a gas cleanliness was latter improved with a turbo molecular pump and ion pump and finally scroll pumps, for all the imperfection of the device it was unlocked purely by earth rotation.

A ring laser senses any reciprocity in the effective optical path length of the counter-rotating laser beams, beat to form a standing wave pattern chapter 2. Initially our signal processing capabilities very limited, we used a New Zealand grown data acquisition system STROBES were crude and plagued by earth loops introducing the mains frequency (50 Hz) which introduced sidebands and their harmonics.

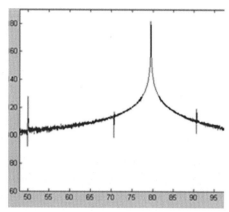

Figure 44, and early Fourier transform (spectrum) of Sagnac frequency in C-I.

Hans was very keen with his new machine (and I struggled to help him) as well struggled to extract useful information from the shapes of these mains sidebands[5], and from their Fourier (frequency) transforms Strobes, was replaced as soon as we could afford it with more fundamental and flexible methods, both of signal acquisition and signal processing where home-built acquisition and MATLAB data processing proved invaluable; also de-trending signal processing methods proved useful. And ultimately through the legendary industry of Ullrich Schreiber in C-II this went to a LAB-VIEW system which eventually grew into a mammoth computer package data acquisition and processing package to control and monitor all aspects of the C-II and G0 UG instruments together including weather station with output(temperature pressure,) and the system also monitored tilt metre readings instrument, Sagnac modulation and backscatter effects on both the counter-circulating beams.

The C-I Canterbury ring system has a nearly square optical cavity with an area of 0.7547 square metres and a perimeter of 3.477 m (this is accurately measured by increasing rf power so as to excite two longitudinal modes of the optical cavity their beams interfering at the detector produces a frequency at 86.22 MHz directly related to the perimeter. Such adjustment of the lasing mode frequency way from the hoped for single mode excitation is assisted by using an isotopic ally enriched He-Ne gas mixture is the medium viz He^{20} and He^{22} to adjust the position in the gain curve for the longitudinal mode with maximum gain of the resided mode. In the larger lasers discussed below. Used a combination of the tricks described above

 In this way the two counter-circulating modes would be set lasing at 633.0 nm. Beams exiting at one mirror through its residual transmission typically 0.25 ppm in later better mirrors with losses of 1 ppm the counter rotating beams could be overlaid using small mirrors and the Sagnac effect of earth rotation is then visible on detecting the interference pattern either with photomultipliers or with solid state detectors. It has measured frequency splitting's. In several respects the C-I was very far from ideal. The zerodur base plate for C-I was damaged during development Zerodur is prone to brittle fracture, and an attempt to bond the blate to the corner boxes ended in disaster, so that its full area of our $1m^2$ plate became unusable. The Bilger design commitment to Viton O-rings to seal the gas system proved unfortunate in that these rings absorbed gas contaminants which were outgassed as impurities into the He-Ne laser gas, (Hans repeatedly found this incredible but we certainly found it a stubborn fact of life.) In those days these O-rings exuded contaminants into the gas system, worst of these was diffusion pump oil. As the project grew we could afford better pumps with reduced potential for gas contamination. Ultimately scroll pumps in all such matters I found myself called on to adjudicate on such matters as, advocates of O-rings and alternatives (not that we had any in C-I) there were many vocal and but strident debates between strident advocates within the group of photo-multiplier on the one hand (where Clive had enormous experience as an amateur astronomer) and solid state light detectors (where Hans also justifiably claimed similar authority) on the other as the detectors to sense the Sagnac signal. I found this particular issue easy to resolved in my judgement, very simply by deciding who was doing the real work and the decision that went in favour of the technicians on the spot who had long-term responsibility for the project. The photomultipliers worked fine although most have now mostly been replaced by solid state detectors given the volume of beam detections that became useful on the larger lasers.

C-I had one enormous engineering problem by it's demand that all mirrors must be aligned perfectly before the system could be sealed and gas filled. This was an extremely arduous operation and external green He-Ne laser whose beams cut through633 nm mirrors could be used for approximate alignment but the final alignment was another matter. Many ideas were floated to improve the situation. With practise the technicians developed tricks to help, for example soft iron strips on the top of the mirror holders themselves place on the zerodur table that could then be manipulated by hand held magnets through the glass covering plates at each corner box, after prior course alignment with a green He-Ne Laser, and a few judicious taps with a screwdriver on the zerodur table while holding the Magnet, the necessary final mrad angular adjustments were possible after assembly and gas filling. All this was a hard exercise for students mainly for this reason heavy technician input proved necessary in this project.

A fuller account of the C-I project is.[6] This whole project was especially satisfying since some New Zealand experts in stable lasers had decided that our project was doomed to failure from the start. For all that our inexperience in Christchurch was un-deniable we were not dismayed by these reports and simply determined to give the project our best shot, which in the event proved successful. I saw no reason to baulk at publicising this fact as widely as the negative comments.

C-II

Given the deficiencies of the C-I machine it was manifest that the road ahead had to involve a quantum leap in engineering and therefore in appropriate financial backing. Professor Bilger made the acquaintance of a German group with related aims to ours. The outcome was a

The C-II ring laser was conceived as a much better well engineered version of C-I, and of similar area. The construction of C-II and its transport to New Zealand was funded by BKG Institut für Angewandte Geodäesie and TUM Technische Universitaet Muenchen, and its siting in New Zealand Cashmere Cavern laboratory of the University of Canterbury the cost of construction was of the order of 2M euro. C-II was built by Carl Zeiss Ltd at Oberkochen, an unusual step for an optical company not known for routinely building lasers. The initial design was by Prof. H. R. Bilger with Dr U. Schreiber and with input from Canterbury technicians and of course Zeiss technicians notably Hieronymus Weber. It has a monolithic construction, as for an aircraft gyro, but within a solid piece of Zerodur 1.2 m x 1.2 m x 0.18 m. Novel engineering problems requiring solution for this project included ultra-high vacuum bonding between metal flanges and Zerodur. Mirrors were wrung on to holding plates which were themselves wrung on to the main body of the instrument; the necessary flat surfaces were prepared by engineering the plates and the main zerodur block all to fractions of a wavelength of light. From the German standpoint C-II was seen as a stepping stone towards an even bigger laser 4m square which would have very serious interest as a tool for geodesy for installation In Germany at the Wetzell research station. This was based on a suggestion by Hans Bilger. Indeed his original suggestion was for a 5m square ring. but in the event it was deemed wise and an intermediate value where precedents like G0 existed was determined to justify design settled that issue. in the event no a plate of Zerodur would not have been unavailable for a 5m ring (Zerodur is manufactured in bulk only at infrequent intervals) (and the plate ultimately used for G was originally made for a telescope mirror).

C-II was installed at Cashmere, Christchurch, New Zealand in a purpose built hut under a solid section of the roof in January 1996 Over the following months it was fully tested and commissioned by Dr Schreiber. C-II was formally opened on 17 October 1996 by Professor Dr Herman Segar director of the BKG Bundesamt für Cartography and Geodäsie Frankfurt. If the truth is to be told the year of commissioning was not as straightforward or incident free as one might have been hoped. The original intention was that C-II would not be altered in any way from the German factory, given the dirty environment of the cashmere cavern as far removed from a lasr gyro clean room facility as could be imagined. However it was discovered at Cashmere laboratory C-II had been filled with an incorrect isotopic gas mixture just one neon isotope (a mistake later attributed to the misslabelling of a gas bottle) so that the lasing was. Fortunately a gas handling facility was asymmetric bequeathed to us by a gyro worker Dr Rodloff Ruediger[7]. When his own gyro project had been shut down. This device was thoughtfully brought over to New Zealand with C-II, down the years since the 'Rodloff' as it was soon dubbed has done us sterling service in more than one of our advanced lasers. On filling with the correct mixture C-II operated satisfactorily.

Much worse happened on two occasions (May 1997 and December 1998) the first at least perhaps from excessive power was applied to the low-loss silica gain tube C-II which

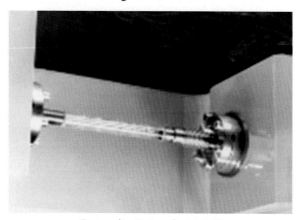

Figure 45, Gain tube in C-II

broke, impacting shards of glass on the mirrors so pitting their surfaces as to make the device totally inoperable. The first loss was attributed to the choice of fused silica for the original tube and its propensity for crystalliation under power cycling.

Such catastrophies mark the history. Of science, certainly our history Professor Bilger had cautioned me years ago with some details of a dramatic and dangerous event that occurred even in the laboratory of the Nobel Laureate Zeeman who described one such in his own very careful research on drag effects for light in moving media.[8]

Canterbury people gave the German suporters of the C-II project enormous credit for the way that this the ultimate disaster was handled. All mirrors from all corners had to be replaced since these had been rung into place by molecular adhesion this was acheived by bringing a trained Zeiss technician Rhuddi (Chapter 8 has a photo), and replacing all mirrors in as clean an environment as could be achieved by a moving tent air filtering systems. The gain tube itself was replaced in pyrex by a glassblower at the university of canterbury in 6mm diameter pyrex as opposed to the previous tube. This brought with it the advantage of introducing relative tube constriction so as to suppressing unwanted transverse lasing modes.The instrument was fully recovered and achieved design goals, part of the commissioning involved stabilising the optical frequency by a difficult procedureachieving new scientific results. In that first year much work went in to stabilising the optical frequency by beating it with an iodine stabilised He-Ne laser while this never became routine procedure the experience gained paved the way for the remarkable successful stabilisation of the G laser in later years. A pressure vessel containing the optical system and removing the atmospheric signal in mirror spacing and laser perimeter was installed in subsequent years this modification likewise was not used routinely, because of the access problems it created but also gave experience for the G system

By the time of the opening event (a photo of the plaque installed at the cavern is later in this chapter) Dr Schreiber was able to lay a firm foundation for another project designed to check the German Dream for a still far better and bigger laser 4metres square which would have enhanced abilities and precision to monitor earth motion. In retrospect C-II data showed a number of the physical effects later noted from other experience such as polar wobble first detected in UG1.

There was immense satisfaction in Germany and New Zealand when Professor Seeger took the occasion of the opening of C-II 17 October 1997, as the time to announce that the (G) project had been given the go ahead by the Bundessampt It was planned to installation (G) in a purpose built cavern at german agency Wetzell Fundamental Research station in the Bavarian Foothills . By this time also on the one hand Professor Bilger had since the C-II project had un-fortunately there had been amased significant disagrements over the building of C_II, and project leader natually became Dr Schreiber who had amassed such significant scientific experience as manager of the C-II project to succeed Prof. Bilger, he had justified this by sheer hard work on the canterbury projects that (G) now naturally became his project as incorparating significant refinements of its design of (C-II) which had been im-plemeted over several years. In New Zealand we welcomed this change of leader over the increasingly difficult personal negotiations.

Today the (G)laser that instrument is the most stable and sucessful accurate such intrument in the world. Several of the instruments we built on the way to G have detected utterly novel physical effects.

Details of the rationale and design of C-II and the mathematical formulae needed to analyse these results in the manner done below have been published elsewhere (see the major reviews of the references) and we summarise here only a few key points. C-II has a square beam path, with side 1.000 m, mirror mountings machined and polished to an accuracy of 10 arc sec, and initially two flat mirrors at the ends of the gain tube section and with two curved mirrors of radius 6.0 m on the other diagonal. The super-mirrors have total losses including transmission of each mirror of 1.2 ppm when new. The laser light path is bored out of a solid piece of Zerodur, 1.2 x 1.2 x 0.18 m, except that in two cut-out sections the path and laser gas are bounded by tubes, one of which functions as a gain tube (under r.f. excitation) and the other of which is available to access beams for research purposes.

The d.c. component from a single beam is used to monitor emergent optical beam power, and so to servo the radio frequency amplitude for driving the plasma in order to keep the beam amplitude at a predetermined level. For mono-mode. Operation with this servo system the required plasma power is typically 0.8 W, the plasma length 10 mm, and the exit beam power (measured for the quantum noise calculation to follow) 10 pW. Any higher power level leads to appreciable excitation of adjacent longitudinal modes The Earth Sagnac rate, nominally 79.40 Hz, was obtained with a measure of pushing and pulling associated with ambient pressure and believed to result from mirror backscatter; details will be published elsewhere. The Sagnac frequency was stable, showing no variation to a precision of 1 mHz when the gain tube was used to vignette the beam, driving the servoed r.f. power to 1.2 W. With a 4 mm gain tube, the laser operated consistently in TEM(0,0). Transverse modes were absent under virtually all conditions. The ring-down time was determined to be 0.20 ms (Fig. 2). This translates into a quality factor of 6.0×10^{11}, and a cavity finesse of 9.4×10^4. We believe this to be a record then for the passive quality factor of an optical resonator. This cavern environment has unusually good thermal and mechanical stability, and its availability has helped to meet the ultimate performance limits of such a device. When combined with the observed output power (10 nW) and mirror transmission of 0.24 ppm, this implies a circulating energy of 1.1 pJ and, with the observed ring-down time, a total power loss (mainly from scatter and transmission) of 5.6 nW. Together, these results imply a quantum noise-induced full width at half maximum of 220 microhertz. This is in good agreement with the observed value of 172 microhertz (Fig. 4). It also agrees well (to within a factor 2) with an Allan variance study of the output frequency in such runs, the defining relationship (Bilger 1984) is that the Allan variance times sample time is half the one-sided power spectral density, itself (within a quantum noise picture) the above-mentioned full width at half maximum power divided by pi. This argument renders C-II of sensitivity 4×10^{-9} rad/s/sqrt(Hz), somewhat greater than that of atomic gyros (Gustavson et al.[12] (1997) who obtained 2×10^{-8} rad/s/sqrt(Hz), Lenef et al. (1997) who obtained 3×10^{-7} rad/s/sqrt(Hz)). These results, and in particular the reduction of drift to a few millihertz, are reassuringly close to the original projections, and augur well for the eventual performance of C-II reaching the parts-per-million level (of the Sagnac frequency) precision as proposed in its original conception. This indicates that the C-II laser system is fulfilling its intended purpose of acting as a prototype and test bed for the much larger (4 m x 4 m) laser G, itself proposed for the detection of short-term fluctuations in Earth rotation. The typical temperature excursion of the C-II pier is less than 10 mK over a few hours.

Figure 46, C-II hut in the cave

Figure 47, C-II on installation day at Cashmere with Morrie, Clive, Hironomous Weber (Zeiss), Ullrich Schreiber, Steven Cooper. It was tested and found operational shortly after arrival

Figure 48, Ulli with C-II at Cashmere.
the C-II facility was formally opened on 17 October 1996 by the director of BKG Professor Dr Herman Seeger.

Figure 49, Plaque over adit entrance.

The C-II ring laser produced new scientific results, reported in many references. Earth tides and sea tides were conspicuous in the rotation record.

We have conclusively detected tele-seismic rotations. recognition vis ring-laser rotation was a novel scientific step. We have hints that we can observe lunar Earth tides.

that is not our overriding limitation to limit further progress at this stage.

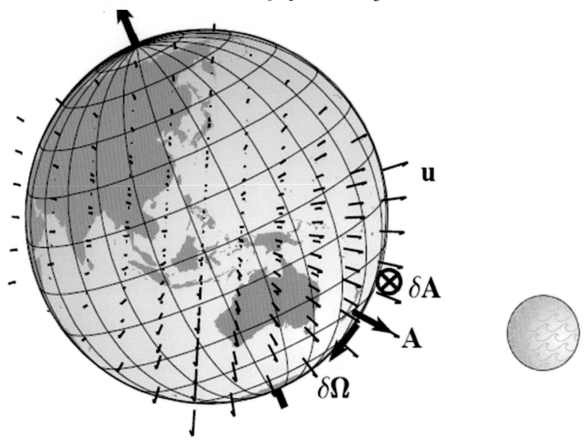

Figure 50. Schematic of the effect of the Moon on solid Earth tides for a horizontal ring laser.[9]

An induced deformation, field **u** moon is indicated; its curl gives a vector which has to be projection on the area vector **A** of a locally horizontal ring in the model used here. However, the associated tilt in this area does have finite projection on the Earth's rotation vector

In retrospect data analysis showed that the. Motion of the rotational axis of the earth later seen in UG1 was evident in a careful reanalysis of the data, local seismic even were conspicuous and the 22 s period mantle waves created by major international quakes of magnitude say 6 Richter or more were routinely observed.

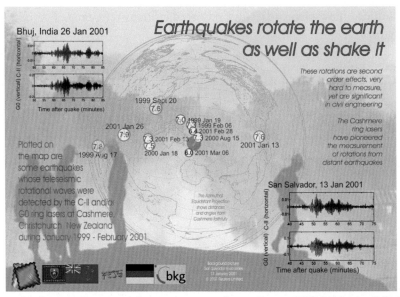

Figure 51, Summarises some tele-seismic events detected over 1999-20001 in C-II data.

including two major serious earthquakes in 2001 in India and el Salvador and was made into a poster as above to demonstrate the point. I always felt is strange that precisely once ever 22 seconds the apparently solid ground at Cashmere could go through this rotational cycle in response the such international events. Papers from. The C-II project have documented earthquake detection, the first study of earth tides was first seriously discussed by Plag et al.[10] and the first major study of a seismic events from C-II data was Panda et al.[11] The project of rotation detection in seismology by ring lasers has since been embraced by professor Heine Igel of UMU (Munich), and a working group on rotational seismology held at the University of Canterbury New Zealand 22-25 September 2013 brought together users of a number of different techniques and very installations. It was advised that UMU now have funding to install a three axis system comparable to G at their observatory Geophysical Observatory Department of Earth and Environmental Sciences Munich University. All of this seems to me to represents a quantum leap in seismology where rotation detection has been difficult.

Figure 52, A German delegation from Deutsche Forschungsgemeinschaft DFG;

Inspecting C-II With Ulli and on the far right Andy Matthews who acted for some years as a liason with German in science. The Cashmere laboratory has retained its useful role as a test-bed on new ideas in particular it has demonstrated that one can build a ring laser of still far larger area.

This size was similar to the area and perimeter of C-I, C-II also had a He-Ne gas mixture for the medium, its two modes lasing at 633.0 nm. In subsequent years there have been a number of significant modifications of the C-II laser.

In December 1998, the C-II system was upgraded with new mirrors and altered cavity geometry as regards the disposition of flat and curved mirrors, in a later upgrade it now has 4 curved mirrors.

In December 1999, the C-II system was upgraded by being encapsulated in a pressure vessel. It then delivered Earth rotation on measurements at parts per million precision. However backscatter effects from residual mirror imperfections which influence the performance in a way which depends very sensitively and in interferometry manner on the interaction between backscattered beams.

C-II is housed 30 m underground in the Cashmere Cavern.

A scanning Fabry-Perot, a Newport SR-130 super-cavity, is used for monitoring the absolute frequency of the C-II beam relative to that of an external Winters iodine-stabilised HeNe laser the collaborative nature of the project is illustrated by the fact that canter bury bought the super-cavity and Germany the stabilised laser. The resultant frequency information is used to servo a piezo on one super-mirror, stabilising the perimeter, and this has significantly reduced the drift in the Sagnac frequency. The residual drift now has an rms variation typically of a few millihertz This drift itself will be significantly reduced by continuing development, including the implementation of pressure compensation, which is now well in progress.

Figure 53, Removal of a corner mirror on C-II

The wood hammer was used to make Rayleigh waves in the substrate which assisted the so combat of molecular adhesion and induce the mirror top off.

Figure 54, A 'clean tent' rudimentary protection for the work area in the cavern environment operated with several air conditioners to help remove dust

Figure 55, The C-II pressure vessel.

G0

Therefore C-I and C-II had proved the principle of 'the canterbury approach' as capable of the aim of building a geophysically interesting instrument, however even then we had not sctually built a 4mx4mHe-Ne ring laser,no instument of such a size having been operated anywhere

The standing component in confidence required to seriously invest in an instrument of that size then was a proof of principle as to wether or not the designs test so far could be scalled up from 1 m² to 4 m². To this end during this years commissioning C-II Ulli Schreiber obtained a very modest support from Munich to build a simple prototype machine G0 which was roughly 4 m²

The only optical table available to us was the concrete walls built by the millitary to support the roof arches on one wall of the cashmere cavern.this instrument (G0) more details are given later in this chapter. This instument was was made to work with some difficulty. This was dueto a calculation error by me on allowable mirror raddi of curvature which while I almost immediately corrected it that correction had not been sufficiently shared within the group. As a result G0 did not lase on its first build. However this was eventually overcome successfully, not before Ulli and I spent a very unhappy Easter together at Exeter University UK while I was on leave debating with a student and a postdocs in NZ who had discovered the correction independently, as to who really had the priority and publication rights over G0. Schreiber could now could return to Munich with clear evidence not only that the fabrication method used to upgrade from C-I to build C-II was successful but that the design could indeed be scaled up to meet the German goal of 4 m² originally set on the recommendation of Professor Bilger after he was disuaded from going to 5m x 5m.

Figure 56, G0 under construction.

G0 was built using cheaper materials, with stainless steel mirror holders- and gas-holding tubes and corner boxes. Mounting was onto a concrete table, which mainly for reasons of cost was chosen as the existing vertical wall of the cavern. This square system has side 3.5 metres, perimeter 14 metres and area 12.25 metres. The Sagnac frequency delivered was 288 Hz. This system has attained 10 parts per million stability in Earth rotation measurements, and with C-II has picked up the rotational effects even of teleseismic waves. Its vertical orientation has the bonus that its seismic information is complementary – it measures Rayleigh (vertically polarised) waves, whereas C-II measures Love (horizontally polarised) waves. The G0 hut is conspicuous in the cover photo.

Figure 57, Clive ascending to the mezzanine floor of G0

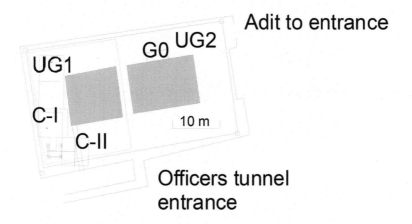

Figure 58, Cavern plan

UG1

This made it possible to contemplate two large lasers whose area included several hundred square metres of rock, these were both built in the course of events. As further exercises in finding to what extent the principles of these lasers could be scaled up and if new effects of interest could be observed. . We dubbed these UG or ultra-G UG1 and UG2 respectively. In each case the gain tube was located near the south cavern wall behind the C-I/C-II huts both projects proved successful, UG2 was however more problematic and UG2 was eventually replaced by a reversion to a configuration more like UG1 called UG3. a television program showing. the entrance drive to the cavern includes views of the interior and the UG2 laser.[13] There are two main scientific challenges in this proposal. One is that single mode gyro operation could eventually prove impossible for the full size proposed for this ring (area of 800 square metres or more). We expect that it will be a significant scientific challenge to overcome this problem in a ring as large as Ultra-G. Two decades ago, this was the main risk in building the C-I ring. One decade ago, it was the main risk in building the G0 ring - a cheap ($20000) test-bed for some design aspects of G. In each case, there was remarkably little real difficulty. Also, we know (as of October 1998) that a ring in Arkansas with an area of 80 square metres and with few special tricks works. Finally we will start with an intermediate area (say 400 square metres) by utilising the side tunnels at the cavern.

The second challenge is that even in the depths of our cave, mechanical stability may prove to be inadequate to guarantee the accuracy and stability of the optical alignment necessary for operation and stability of such an enormous laser.

Figure 59, Early days of UG1 Rob thirkettle (Canterbury) , Dima Schabilin (Dubna Russia, , Felix Puls (Tech. Univ. Munich) Bob Dunn (Hendrix College),

Figure 60, Rob and the gain tube for UG1/2 by the south wall of the cavern

The dimensions of UG1 are 21.0 metres by 17.5 metres, so giving a perimeter of 77.0 m and an area of 367 square metres-a world records, eclipsed by UG2 and so these lasers represented uncharted territory for research. UG1 is a relatively conservative increase from a successful triangular ring laser gyroscope built by Professor Robert Dunn at Hendrix College, Conway, USA. Who participated in the building of UG1 (an early version with the beam in air) during August/September 2000. The laser gas would later be enclosed in a stainless steel and evacuated tube skirting the perimeter walls of the cave.

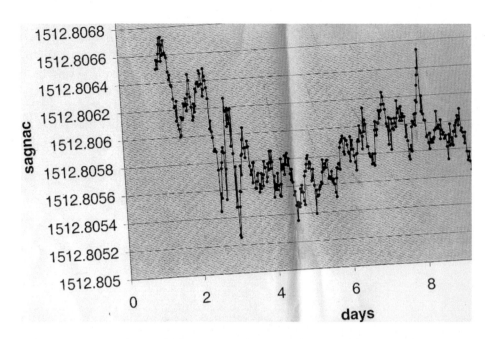

Figure 61. The daily signature in this record from the early days of UG1 is dominated by lunar induced wobbles' of the rotation axis of the earth, the moons gravity acting through the equatorial bulges on the earth [as if the earth is an out-of-balance car tyre] to set up a daily oscillation in which the word pole describes a circular of diameter of the order of 60 cm.[14]

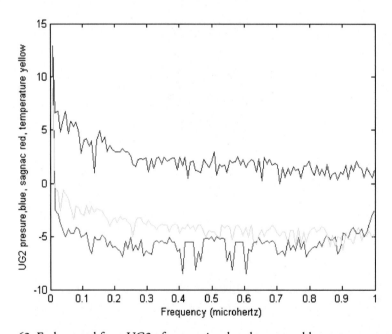

Figure 62. Early record from UG2 of sagnac signal, red, pressure blue, temperature, yellow

UG2

This machine went all the way around the cavern and so had record perimeter rectangle aproximately 21 m ×39.7 m and area 834 m²

First light for UG2 was on 13 december 2004 It proved to be severely affected by back scatter effects and was later abandoned for a ring UG3 similar to UG1. Still given the relative budges of G and UG2 $USA19 000 000 $NZ100 000UG2 was a reasonable experiment and all useful experience for the collaboration

Figure 63, North-west arm G0 hut on right adit to cave entrance in distance

Figure 64, Ulli with John Mander, seismic engineer (from Canterbury now at College Station, Texas)

PR1

Two geo-sensors of relatively small area were built in Germany one was located at the University of Canterbury where it sensed the rotations of the physics building the other was sent to Piñon flats USA, see below PR1.

Ulli has had several side projects for example A commercial fibre optic gyro was used to sense the building resonances of the Auckland Sky Tower and also an involvement in the control system for the clocks of Mecca Royal Hotel Clock Tower 601m whose clock faces are 43m square.

G

This is a horizontal planar ring at the Fundamental Research station, Wettzell Bavaria, latitude 49.1441 N . the cost of this project was of the order of $17M. Many introductory explanations regarding G are given in earlier parts of this chapter.

Figure 65, As announced in New Zealand our German partners proceeded with their aims of building a well engineered 4m² laser G. The Zerodur table was Once destined for an astronomic telescope, such large pieces of zerodur are cast only at at frequent intervals at Schott Mainz four zerodur arms bolted to it supported the mirrors and define the lasing cavities and their connecting tubes.

Figure 66, Side view of G

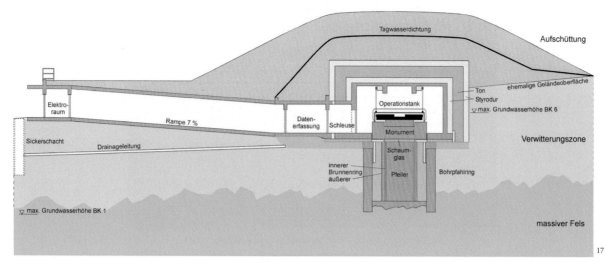

Figre 67, Side plan of laboratory complex

Figure 68, Installation day

Figure 69, Preparation at Schott Mainz.[16]

Figure 70, Top of the 'monument or G-support pier

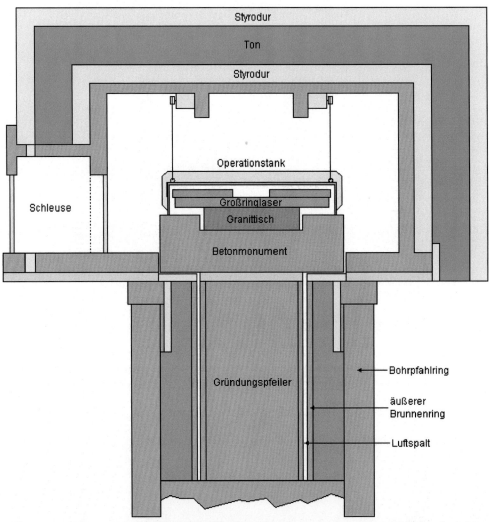

Figure 71, The concrete pier was 90 tone firmly bonding the instrument to the motion of the earth and insulation[17].

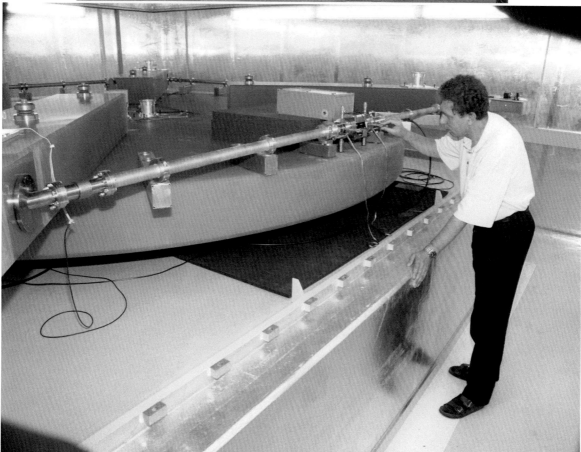

Figures 72,73, Side views with ghetter tubes.

Figure 74, Ulli at G

At the opening of the G laser Opening

Figure 75, B. G. Wybourne Representing New Zealand, then based in Institute of Physics, Nicolas Copernicus University, Torun Poland. H Weber (Zeiss Obercochen), W Schlueter (Wetzzel) M Schneider(T.U. Munich)
(copyright Geodetic Observatory Wettzell)

Brians reaction to seeing the G ring

Ulli once sent around this a document on Brians visit to G. While living and working in Poland, Brian frequently visited some colleagues at the "Max Planck Institute" in Garching a suburb of Munich. Since he was often passing reasonably near our Geodetic Observatory at Wettzell, a small village in the Bavarian Forest close to the border of the Czech Republic, several times he took the opportunity to visit (or should I say inspect) the ring laser project. We often spent hours in discussing the status of the project, the next goals and the obvious milestones. His firm belief in the ring laser technology was very vital for our project in many ways. The dream of ultra high resolution sensors for angle or rotation rate measurements was widely oversold in the United States during the 80th of the last century. Since many developments did not perform to their expectations by a long shot, funding for such projects became extremely difficult in the USA and also caused wide criticism in Europe. Brian's support and vigorous but well measured optimism has always been noticed and remembered when the funding of the C-II and G ring laser was discussed. It is always good to be cautious about those people who actually do a project all they want is money. Brian's position was different. He had experience in ring lasers but he did not have a direct involvement in the proposals under discussion. At the end of 1996 when C-II had 'first light' and showed signs of the Sagnac signal, Brian immediately hoped on a bus and toured all the way from Torun in Poland to the Carl Zeiss Company in Oberkochen (a small town in the south-west of Germany) to see the then largest ring laser. After 22 hours of travel cramped into an overfilled bus he arrived at Oberkochen, muttering that never in his life he would do such a horrible thing again. However, when he saw C-II, all the aching bones were immediately forgotten and he exclaimed: "When the really large ring G is done, I have to see it – even if it means walking from Warzaw to Oberkochen. Fortunately he did not have to walk that far. In October 2001 he represented the New Zealand group at the opening of the G ring laser at the observatory in Wettzell. His talk was surely the most important one on that day Speech by Brian Wybourne - Talk given at Wettzell 5 October 2001[18] 'As noted in the last reference This talk was also reprinted by Lydia Smentek, and in that version photographs of the Polish gravestones of both Mickelson and Wybourne himself are included'[19]. On Monday I shall be driving to Torun, in Poland, the birthplace of Copernicus. On the way to Torun I will pass the small town of Strzelno, the town from which the Michelson family emigrated to the USA with their two-year old son, Albert. Albert Michelson came synonymous with the Michelson interferometer. In 1894 Michelson spoke at the dedication of the Ryerson Physical Laboratory, University of Chicago saying "The more important fundamental laws and facts of physical science have all been discovered, and these are so firmly established that the possibility of their ever being supplanted in consequence of new discoveries is exceedingly remote . Our future discoveries must be looked for in the sixth place of decimals."

Michelson's interferometer started as a table top experiment and grew to the large Clearwater interferometer. Michelson brought to physics a new scale of precision and changed our perspective on space-time. He can hardly have conceived, at the time, the subsequent developments of his interferometer which is today realised in Very Long Baseline radio telescopes such as that between Bonn and Torun or the incredible interferometrically coupled quartet of 8.2metre optical telescopes operated by the European Southern Observatory in Chile.

Today we dedicate a new instrument that gives us hitherto undreamed of precision - not six decimal places but many more. Michelson was a dreamer and envisaged ever bigger interferometers.

The current instrument had its origins in New Zealand in the 1980's when Hans Bilger (University of Oklahoma at Stillwater) and Geoff Stedman (University of Canterbury) collaborated in Christchurch in the construction of C-I on a 1m x 1m slab of Zerodur. The initial development involved much learning, frustration and at times excitement. It was the coming together of two physicists and two dedicated technicians, Morrie Poulton and Clive Rowe that made the impossible possible. I recall the excitement when the first arm of CI lased then the second sometime later, then after a coffee break the Chinese student, Ziyuan Li, said 'I saw fringes!' The impossible was achieved. CI involved a steep learning curve.

C-I was followed by the development of the NZ-German collaboration led by Manfred Schneider and Ulrich Schreiber. This led to the remarkable C-II that brought on board Carl-Zeiss and Schott with all of their incredible skills

With the success of C-I it was natural to take the next step which today we celebrate. A splendid advance of science from technology. Michelson may have been impressed for, as with C-I, he started with a table top interferometer which grew to the Clearwater experiment which was no longer possible to place on a table top and so with the Grosse Ring.

Already C-II and G have simultaneously observed seismological events that involved collaboration over vast distances. But this is just the beginning. We have entered an entirely new range of precision. In the future we expect such ring lasers to increase our understanding of the dynamics of Planet Earth. I am confident that further developments will lead to observations of such esoteric things as CP violation, gravitational waves, terrestrial observation of the Lense-Thirring effect and of things as yet undreamed of.[20]

The Grosse Ring demonstrates science as international and the essential results of collaboration with no borders or boundaries. To you all who have contributed so much I offer my congratulations. This is the beginning of the project - the end is unknown. We have commenced a journey that has no end. Thank you!

Figure 76, audience at G opening
(copyright Geodetic Observatory Wettzell)

In September 2013 the absolute frequency of G was stabilised against a maser standardfrequency counting standardwith the resulting Allen deviation plot

At the time of writing G has gone through several major upgrades. during its operation G has experiences several upgrades of mirrors of the installation of getters to keep the gas clean, very recently of determining its absolute optical frequency by frequency counting from the microwave. Its most impressive accomplishment to have detected the chandler wobble of the axis of rotation of the earth discovered by Seth Carlo Chandler in 1891[21]. The motion of the north pole is buy up to 30' over a period of 433 days, the ring laser observation is docuｍemted at.[22] And the essential results are reproduced in the figure below.

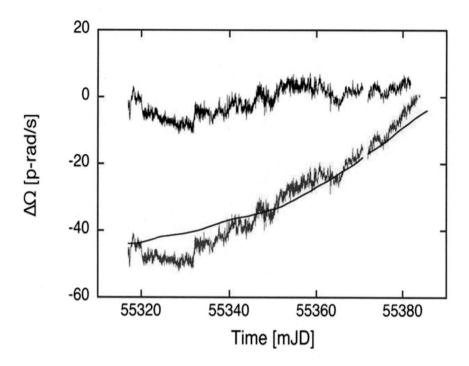

Figure 77, The red line shows the ring laser observations with local tilts and diurnal polar motion removed. The blue line represents the polar motion signal caused by the Annual and the Chandler wobble, measured by the VLBI radio telescope network. Previous figure the black data is obtained when this Chandler signal is removed from the ring laser observations. The remaining offset in the data is caused by backscatter coupling in the ring laser. More recently the absolute optical frequency of G has been stabilised against a frequency counting standard.

It is important to note, that VLBI only generates 2 data points per week, while the ring laser provides a measurement every 30 minutes.

PR2

PIÑON FLATS

This seismic observatory is operated by the Institute of Geophysics and Planetary Physics of the University of California, San It has a PR-1 type ring laser has been employed as a field instrument for seismic studies at piñon Flats[23]

GRAN SASSO

The University of Pisa in Italy has in collaboration with the VIRGO Gravitational Wave detector project endeavoured to measure rotations with ring lasers to determine their effects on that experiment.

A proposed installation is a tetrahedral ring in a nuclear physics laboratory in a side tunnel to a motorway at Grand Sasho.

Figure 78 (a) and (b)
(a) At top (GINGERino) is a square ring 3.6m side, assembled inside the Gran Sasso. Underground laboratory in a motorway tunnel excavation'[24].
(b) GP2 is the prototype in Pisa 1.6 m on a side. Developed to study suitable control strategy in order to keep Sagnac signal constant (1 part il 10^{10}) the technician is Filippo Bosi.

Figure 79, Meeting with visiting italian collaborator Clive Rowe, Ulli Schreiber , Angela di Virgilio (Pisa) bob Hurst, John-Paul Wells at Physics Department University of Canterbury New Zealand.

Several more ambition projects have been discussed.

MUNICH

In 2013 we were advised that Professor Heiner Igel from the Ludwig-Maximillian-University of Munich received a 2.5 M€ grant from the European Research Council (ERC) to construct 4 large stainless steel ring lasers 12m on a side of the quality of G, arranged in the shape of a tetrahedron, buried with the tip 14 m under a plot of grass at the Geophysical Observatory in Fürstenfeldbruck about 30 km west of Munich. Such a machine will certainly revolutionizes the field of rotational seismology.

To build a triple axis ring laser device dubbed **ROMY** which stands **RO**tational ground **M**otions a new observable in seismolog**Y** now under construction in a field near Fürstenfeldbruck west of Munick.

Figure 80 The First Large 3D Ring Laser Structure for Seismology and Geodesy

It is a full 3-dimensional rotation sensor. the ring lasers are rigidly tied together to the same mechanical reference. early results are at {https://www.geophysik.uni-muenchen.de/ROMY/pages/news.html}

MCR

An unfulfilled dream to my mine is to replicate with lasers the Achievement of Mickelson in measuring earth rotation using spectral lamps rather than lasers in an interferometer of gargantuan dimensions. A rectangle of water pipes[25]

was laid out in clearing Illinois, in 1925 a rectangle 612m x 339 m area 207468 m² and perimeter 1294 m he observed the contribution of the earth rotation to the fringe shift with the aid of a small rectangular interferometer in one corner[26]

As Larmour said to Lodge's on his own experiment (chapter 1) "It is suggested that you are going to reverse the rotation of the earth in order to get an interference effect around your circuit!" Michelson solved this problem elegantly in 1925 by utilizing a smaller interferometer built into the larger one for calibration of the fringe position.

Einstein said: "My admiration for Michelson's experiment is for the ingenious method to compare the location of the interference pattern with the location of the image of the light source. In this way he overcomes the difficulty that we are not able to change the direction of the earth's rotation." [27]

Today I believe his would be far more easily done using the kind of helium neon lasers and mirrors now available and used in the project listed in this Chapter. We are approaching the century of Michelson's work, 2025 and a project to replicate it would be a fitting target for experimental science[28].

The idea of an MCR as a fitting scientific anniversary celebration of the achievements of Michelson had been mooted by us earlier[29]. To be sure It may be argued against such a project that it would little more than was argued against such a project as it was against Michelson's project that it would prove only that the earth rotated a fact of which we are well assured already. Of course I was reminded that the Sagnac signal in such a device as MCR if built in a trench in open country would have to contend with every source of drift under the sun but for all that. Still given the surprises the above projects have yielded I would have higher hopes of a novel outcome of a Michelson Centennial Ring.

Figure 81, supporters of a Michelson Centennial Ring, Hans Bilger with grandson Hans

Clearly Hans Bilger was another person who shared my dream, he sent me this photo as evidence Their headgear was apparently a spinoff from a local shopping mall visit by Hans and his grandson Hans very appropriately in Illinois.

During Hans last visit to Canterbury in the early 2000's this was in fact seriously discussed with the head of department Phil Butler as a possible show piece project for the collaborations, but it would require a new laboratory involving trenches on the Canterbury plains somewhere such techniques are well accepted in gravitational wave detection, admittedly Hans was admittedly opposed to this idea preferring a purpose built tunnel in the port hills, however the war is over.

And by that stage neither of us had the health to initiate and maintain such an effort, some argued strongly that a 100m x 100m square pilot project with the possibility of extension was arguably a good first step, but we did not have any source of the funding necessary to commence such a project so that idea has been a casualty of history it would seem. I fear that the Michelson centenary will pass unchallenged by lasers gyro's.

LT laser

a triangular laser is in an advanced stage of planning, although it's construction has been halted by the Christchurch earthquakes which to date have closed the cavern as well as destroying many historic buildings and residential homes

Ring Laser Gyro Measurement of Absolute Earth Rotation Rate

Since 2011 a Marsden project has been running with the goal of measuring absolute rotation rate of the Earth, to a target accuracy of 1 part per billion (ppb). This is sufficient to resolve variations in LOD (length of day, conventionally the departure from nominal 86400 s) at the level of 0.1.ms. Variations in LOD occur because of (smallish) angular momentum exchange between the solid Earth and the oceans and atmosphere. The experiment has the potential to measure an expected discrepancy between LOD[measured with respect to an inertial frame, and LOD measured astronomically.

Relativistic precessions (Lense-Thirring or frame-dragging effect see chapter 5, and possibly various compelling effects geodetic and Thomas precessions) at the level of a few ppb. Despite what we believe is a viable experimental design, experimental progress has been minimal because of quake-related damage to the Cashmere Cavern. Their has been discussion about joining an experimental project in the Gran Sasso National Laboratory in central Italy as described above, where the Pisa group have similar aims with a rival proposal[30.] If built, our ring laser design will be triangular, about 5-6 m on a side, one side level along the ground, the remaining corner high on a wall. Its plane will be normal to the Earth's rotation axis. Meanwhile we have made theoretical progress in understanding and correcting for the effects of backscatter in ring laser gyros, usually the largest source of systematic error in measured rotation rates. We have located a non-traditional material, carbon-fibre-reinforced plastic, that has rigidity comparable to steel and thermal expansion coefficient competitive with invar.

In conclusion

We have had to battle a mountain of scepticism to enable each of these instruments to be built. For example when the MCR (proposal was aired I was continually told that the scientific objectives of an project MCR were ill-defined.

What I would say to that is that "Nobody Absolutely Nobody" 'said to me before UG1 was built'. That UG1 machine would then be able to measure the daily effects of the moon on the axis of rotation of the earth 'As said by Albert A. Michelson "My greatest inspiration is a challenge to attempt the impossible." 'What is scientific research if not a leap into the unknown to see what can be found but once that had been achieved and understood, and a superior instrument like G was available the Chandler wobble detection was fairly in the sights of such an instrument and measurement technique[31].

note on massive particle interferometers

Because of the $\hbar$ and the particle-mass-dependence of the de Broglie wavelength. Interferometry using massive particles is very attractive with high potential sensitivity guaranteed by the founding principles of quantum mechanics (chapter 6). Some pioneering work with superconducting liquid helium and atomic interferometry s illustrated this, however none of these devices has as yet produced the results of geophysical interest as those discussed above from our large ring lasers. It is a pity therefore when the practitioners here tend to denigrate laser gyros.

Packards[32] talks of 'unseating', ring lasers with his liquid helium gyro's.

Packard said "These (liquid helium devices) may be a better mousetrap, but we don't know if there are any mice to be caught." 'It is really too early to predict what other uses there may be in future.'

That was precisely my policy in proposing UG1 and UG2. They had potential before G to be the best mousetraps of their day, and the successes of UG1 and G have vindicated that expectation which liquid helium and atomic gyros have not yet matched.

Kassevitch's group says in a recent paper lasers are 'noisy'[I presume that means limited by nothing but quantum noise] I say show me the experiment without a limit of some kind that has has not proved a practical limitation to experiments, in our case backscatter on mirror defects is the one thing we would love to get rid of, but it is not quantum noise is not our overriding limitation.

Endnotes

[1]   G.E. Stedman R.B.Hurst U.K. Schreiber Opt Communications **129** 124-129 (2007).

[2]   G.E. Stedman,, R.B. Hurst. K.U. Schreiber Optics Communications **279** 124–129 (2007)

[3]   http://www.rotational-seismology.org/events/workshops/2013
http://www.springerlink.com/content/1383-4649/16/4/

[4]   R. B. Hurst, G. E. Stedman, K. U. Schreiber, R. J. Thirkettle, R. D. Graham,
N. Rabeendran, and J.-P. R. Wells, Experiments with an 834 m2 ring laser interferometer JOURNAL OF APPLIED PHYSICS 105, 113115 (2009)

[5]   G E Stedman and H R Bilger, Ring laser, an ultrahigh resolution detector of optical nonreciprocities, Digital Signal Processing, **2** 105--109 (1992).

[6]   GEN A

[7]   Inst füer Flugfuehrung Deutsche Forschungsanstalt fuer ALuft- und Raumfahrt Braunschweig

[8]   P. Zeeman , W. de Groot, A Snethlage and G.TC. Diebetz. Proc R. Acad. Sci. (Amsterdam) **22** 1402-1411 (1221) And references therein.

[9]   GEN A

[10]   V Rautenberg, H-P Plag, M Burns, G E Stedman and H-U Jüttner, Tidally induced Sagnac signal in a ring laser Geophys. Res. Lett. **24** 893-896 (1997).

[11]   Pancha A, Webb T H, Stedman G E, McLeod D P, Schreiber K U,
Ring laser detection of rotations from teleseismic waves, Geophys. Res. Lett. 27 3553-3556 (2000).

[12]   Gustavson et al., Phys. Rev. Lett. 78, 2046-9 (1997)

[13]   http://www.ecasttv.co.nz/program_detail.php?program_id=1921&channel_id=84&group_id=

[14]   K U Schreiber A Velikoseltsev T Klugel M Rothcher, G.E.Stedman DLWiltshire
Journal of geophysical research **109** B06405 (2004)

[15]   GEN I

[16]   http://www.schott.com/magazine/english/info95/si095_04_laser.html

[17]   GEN I.

[18]   Brian Garner Wybourne Memories and Memoirs adam marszalek Totun(2005)

[19]   (GES was unable to attend after a severe stroke that year.)

[20]   G.E. Stedman K.U. Schreiber and H.R. Bilger Quantumgrav. 20 5527-2540 (2003)

[20]   J. Belfi1 N. Beverini1 G. Carelli1 A. Di Virgilio2

[21]   http://en.wikipedia.org/wiki/Chandler_wobble

[22]   K. U. Schreiber1,2,*, T. Klügel2, J.-P. R. Wells3, R. B. Hurst3, and A. Gebauer1 Phys. Rev. Lett. **107**, 173904 (2011)

[23]   http://eqinfo.ucsd.edu/deployments/saa2.php

[24]   http://en.wikipedia.org/wiki/Laboratori_Nazionali_del_Gran_Sasso

[25]   "http://www.usna.edu/LibExhibits/collections/michelson/"

[26]   A.A.Michelson A.A.Gale and H.G.Person Astrophysics Journal **61** 137-245 (1925)

[27]   GEN B.

[28]   The MCR suggestion first appeared in the paper Stedman Schreiber and Bilger Classical and Quantum gravitation **20** 2527-2540 (2003)

[29]   G. E. Stedman K.U. Schreiber and H. R. Bilger the detectability of the Lense-Thirring field from rotating laboratory masses using ring laser gyroscope interferometers}, Classical & Quantum Gravity **20** 2527-2540 (2003)

[30]   F. Bosii et al. Physical Review D **84** 122002 (2011)

[31]   K. U. Schreiber, T. Kl¨üngel, J.-P. R. Wells, R. B. Hurst, and A. Gebauer Phy s, Rev, Letters **107** 17390 (20011)

[32]   http://berkeley.edu/news/media/releases/97legacy/gyroscope.html

8 Personalities

The personal element is as inescapable in Science as in all human endeavor this project certainly illustrates this.

Alpers cautionary Tale

In the words of a celebrated Christchurch author the Danish Judge Oscar Alpers. The city of Christchurch has always been a happy hunting-ground for eccentrics[1].These were the days when Latimer square was dominated by con man Arthur Bently Worthington's; Temple of Truth[2].

Even in my student days (says Alpers) I came to know some of the cranks. In my freshman's year I went one Sunday night with a number of other undergraduates to a lecture. The advertised synopsis attracted us. Bill-posters and newspaper advertisements gave out the subject: "The Earth is Flat-after all." This was really more than we could stand. As undergraduates in our first year, we flattered ourselves that in addition to being omniscient, we knew a little mathematical geography, and so we attended the lecture with the deliberate intention of "rocking" the meeting. But we didn't. A very old man with a most venerable and kindly appearance came on to the platform, and, without any chairman or other person to introduce him, opened his theme. He had a very attractive speaking voice. There was something magnetic about him: he held his audience, undergraduates and all, spell-bound. With many learned citations from the ancients, with much abstruse mathematical calculation, he demonstrated that in spite of the evidence of the theodolite, the phenomena of the visible horizon, the shadow of the earth seen in eclipse all the physicists and astronomers were wrong, and the earth was indeed as flat as the ancients had conceived it to be. Those were days when public meetings sometimes became riotous, and when stale eggs and dead rats were occasional weapons in popular controversy; and yet that audience, undergraduates and all, let the old man prattle on without serious interruption, and I like to record it as a merit in my fellow-collegians that it was a particularly unruly undergraduate who at the close of the lecture, in grave and perfectly decorous terms, proposed a courteous vote of thanks. the situation was full of humor, but it was also instinct with pathos; the audience fortunately appreciated the pathos quite as much as the humor. *For the irony of it all was that the lecturer was a surveyor who, with his own theodolite, had surveyed much of the Canterbury Plains.* When he was an old man, senile decay overtook the tired brain, and it assumed the ironical form of this obsession.

One must always be on one's guard in science If I might venture to add a rider to such a story. I know a Christchurch school teacher in science who decided to introduce his little darlings to the scientific method, so he proposed to his class that the earth is flat. And invited them to disprove it. He made a few useful suggestions to guide them like ringing up their friends in USA UK etc. in the middle of our night and asking them where the sun was. At the next parent teachers meeting he got accosted by one anxious parent: so you're the man who believes the earth is flat, the explanations grew inevitably a little heated and protracted. well such science education can be in New Zealand; I can only assure the reader the to my personal knowledge the Canterbury plains are still mainly flat despite or because of the best efforts of Alper's surveyor. And the flat earth society. Amongst other things this Chapter will explain how we got some international help on such matters. [3]

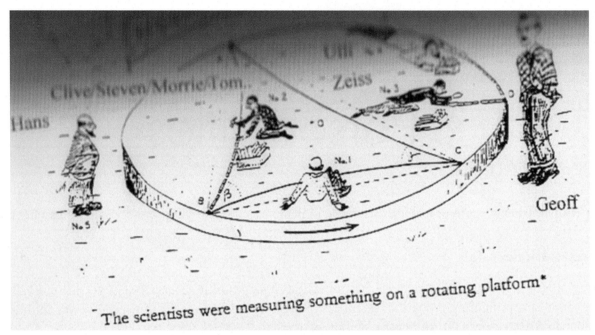

The scientists were measuring something on a rotating platform*

Figure 82, With apologies to Gamow group cartoon.

Figure 83, An early photo of the C-I collaboration, Peter Wells (MSc PhD student), Brian Wybourne(head of Department), Hans Bilger initiator Frank Kowalski who first got C-I going in the Physics Department Geoff Stedman Clive Rowe Morrie Poulton technicians

Figure 84, A recent photo of the Canterbury group Geoff John-Paul Wells Bob Hurst, Rob Thirkettle Ullrich Schreiber, at the time of the 3IWGRS_programme Christchurch September 2013.

Dr Archie Ross

Figure 85, who first taught Geoff operational relativity (which stimulated related questions chapter 4) Archie with Rod Syme constructed New Zealand's first He-Ne laser from a USA kitset in our Physics Department though some mirror issues needed resolution.

Professor Hans Bilger

Figure 86, The father of large ring lasers projects is Hans Bilger a native then with his wife Edda of Maggia a village in Germany near the French and Switzerland borders and so well used to the problems of Post War Germany, Hans did a PhD in Physics at Basle University under Professor Firtz, Hans was, for some time at Raytheon research division of research associates Montpelier France then senior research fellow at Californians to technology. He was called an expert witness may minimise a court case on aspects of noise theory and expert in semiconductors. Hans was particularly proud of his early work on the Fano factor in Ge work[4] which I had attracted a very positive comment being described as a classic paper in a Wiley book by van der Zeil's on noise measurements he also performed the first measurement of noise in a double injection diode he emigrated to USA eventually getting a teaching post at Oklahoma University in Stilwater OK, where he remained until retirement in during this time he developed his interest in ring lasers with summer vacation spells at Raytheon where the emphasis was on military and commercial aviation hardware; for commercial inertial navigation. One major interest at this stage was an ultra large ring laser in a project at Seiler Air Force Base, Air Force Academy Colorado USA this device was planned to utilise a stabilised platform with area of 58 m², is was deemed to require $196k worth of Zerodur (An alumino-silicate glass-ceramic with unusually low thermal expansion), manufactured by Scott, Mainz, (I rather believe this little item was included at Han's urging) and with many millions of dollars invested in vibration free corner supports. It aimed to detect the frequency of earth rotation of precission of one part in 10^{-9}. The cancellation of this project was announced in a letter from the commander of the US air force base dated January 1986 i.e. shortly before Hans initial letter to Geoff. and it was undoubtedly this letter which turned Hans mind to searching for other possible scientific groups who could collaborate on a science rather than a military project given the availability if only in the military area at that time of mirrors of previously undreamed of reflectance's up to 99.9999%.

Hans had otherwise been unable to find backers in the USA for his interest in doing fundamental physics with ring lasers, whose military development had now passed critical initial stages and which were now at the heart of avionic technology for inertial guidance of aero planes and missiles of all kinds. As such they had made mechanical gyroscopes of the older kind obsolete. it is fair to say that mechanical gyroscope has made a major come-back for some applications in a much revised form used silicon fabrication technologies (chapter 9 briefly comments on this technology). The emergence of the ring laser for guidance purposes was fuelled by dramatic developments in the mirror manufacturing industry both through engineering by which highly smoothed surfaces are by perfecting techniques of coating them with perhaps 22 or so thin film coatings to enhance their reflectance for a given wavelength of light (in our case the helium neon laser wavelength of 633 nm at a definite angle (for us 45°). This advance was the chief step making our large ring laser gyros possible.[5] It was our dream to capitalize more on these considerable advances in technology for the benefit of studies in fundamental science rather that merely the fuel the growth of the world industry of superior armaments. The Bilgers Catholic upbringing did not well suit the military orientation of the application of ring lasers in controlling the flight of missiles and the warhead's. Although that did not stop us all ending up in a military bunker (chapter 3) for an optics laboratory which had no better use once the Pacific phase of World War II was over. Raytheon for whom Bilger worked in summer slots had military interest for example in the Patriot missile guidance systems.

Hans opened a casual correspondence with a letter to me out of the blue with a letter dated 13 January 1982. it started by explaining much of the above background and he mentioned having read 'an old is a letter of mine' in American Journal of Physics[6] ring lasers are closed path devices as probably the high precision devices for testing the equality of light speeds on oppositely directed paths around a polygon (chapter 4 gives more of this letter, he continued).

I might mentioned that a paper of ours[7] has been greatly improved by another student of mine W K Stowell in the mean time ring lasers has have improved considerably, since they are application as gyros pushed the industry. As an example, see the paper of the group that I am participating as consultant and summer guests[8]. Where frequency fluctuations below 0.1 Hz had reached the time several hundred seconds. Indeed the Claim of having reached the quantum limit indeed now established. Experiments come now into reach which in the past were experimentally not available.

Note that the relative fluctuations of these oscillations are $\Delta f/\Delta f \sim 10^{-18}$. Several orders of magnitude of smaller the and the next new in line the Mössbauer effect. We are currently investigating possible research topics with such great resolution lasers in How far away are, in your opinion, general relativistic experiments in the laboratory (detector and added source in the lab). According to my estimate, (a to be a simple Lense-Thirring experiment with a giant rotating mass is still out of reach by several orders of magnitude, Hans's interest at that stage was in a rotor experiment as with he 600 tonnes 133 rpm hydro-electric generator at Quebec . The avionic and guidance industries had pushed for state-of-the-art and devices of area the order of dm^2 Hans speculated on the possibility of laboratory tests of general relativity for example Lense Thirring experiments or dragging of local inertial frames from earth rotation [as of 2013 this is becoming more feasible, because to innovative ring designs, a program which has been much delayed by earthquake damage in Christchurch]. Hans also floated a number of these possible applications to fundamental science recognising my potential interest in such. for example do you think that it may be possible to find a gravitational Aharanov Bohm effects from the moon or sun these questions made general present interest I would greatly appreciate your thoughts on the subject. Hans also noted very prophetically our proximity to seismically active circum-pacific belt and geophysical interest in studying possible rotations anticipating that Fourier spectra would distinguish such from other effects a physicist might have articular interest in gravitational waves preferred frame effects and the like. He then summarized: this brings me to the head of the letter since an earth bound ring laser will sense any the rotations of its base the effects due to the direction of their can be sought out from the response of several stations around the earth the effects of frame and three gravitational waves and preferred frame are a characterized by different arriving times could be distilled out of the combined output of the ring laser at Christchurch New Zealand be a very interesting proposition together with a third station at Nairobi Kenya latitude 1° 1" merely equatorial would form the almost equidistant stepped on great circle of the world. On 17 May 1982 Hans clearly asked whether there was a group in New Zealand or Australia where experimental ring laser research could be pursued. Clearly in these letters Hans is floating somewhat wildly a wide variety of possibilities of quite distantly related and ambitious experiments which might require new ideas". His letters arrived at a serendipitous moment on several counts. First I was in fact on the point of giving a seminar on ring interferometry because I had been impressed by progress in neutron interferometry and superconducting SQUID interferometers, I mentioned Hans interest as part of that presentation and was pleased at an interested reaction. I replied to Hans on the first of February to say that I was delighted in his interest

I had no idea frequency resolution had reached this level of precision. Secondly as Hans had detected I had been following such far out ideas myself for several years for example I had been following some ideas on the predictability of QED the effects by optical experiments from a group at the University of Sussex that I later found to be worthless. However Hans's letter coincided with two items givening it serious interest. The third serendipitous aspect of Han's letter was that it coincided with a wish by the new head of department Brian Wybourne for a new project as a goal for the workshops, they had just successfully construction of a 1m diameter telescope for the astronomical observatory at Mt John, and Brian was searching for something which would be a flagship project for the Department and our workshop staff. Indeed I was totally surprised at the urgency with which Brian followed up the original suggestion from Hans. The upshot was that Hans Bilger was invited to take up an Erskine fellowship in the Department so that Han's first visit to Canterbury was from 1 May 1988 to 15 August 1988.

Discussions in that time focussed on building the laser later called **C-I** (see chapter 7.) It was judged possible to build an instrument much greater in size than the typical aircraft gyro's which was stable enough not to be overwhelmed by quantum noise or flicker floor noise, could hopefully operate in single mode by starving the others mode by RF control also that it could be unlocked by earth rotations alone. (No doubt these hopes were on the table at the time of the Seiler-lab design also though I have no details to confirm this.) It could be excited by Radio Frequency built on a small budget locally and had the potential to produce novel scientific results. During their time with us Hans and Edda regaled us with stories of their upbringing during which they well remembered the ravages of the closing phases of WWII in Germany. The liberating troops from USA offered the starving population flour made from maize which made bread unpalatable to the German palate, the German children decided that the USA having won the war were trying to poison the Germans.

Brian Wybourne pleasant summary of his interactions with Hans and Edda Bilger and Geoff over the ring laser project are published by Lydia Smentek[9].

The C-I project commenced in earnest about 1990. It a was successful, it had attracted wide interest and in particular brought us in touch with a German group with similar aims. This group eventually decided to proceed with a far better machine, C-II (chapter 7). Unfortunately Professor Bilger amassed significant disagreements over the building of C-II and decided to lay public claim to the projects. We in the New Zealand group were singularly unimpressed on discovering that Professor Bilger had wroten a harsh letter to the magazine Science[10] about the project management and credits, And publicly claiming that C-II was entirely his design. The fact was that the other parties involved and in particular Dr Schreiber had by now amassed such significant scientific experience and by dint of sheer hard work C-II and later(G) now naturally became his projects. He and also workers at Zeiss had contributed several later significant refinements towards the design. and these projects had continued essentially without Professor Bilgers input. I knew little of these problems before this Science paper appeared.At one later stage Prof Bilger sent me a e-mail that for 10 years Dr Schreiber had chosen not to collaborate with me; It had been very evident that for the same time Bilger had not collaborated with me Zeiss (who fabricated **C-II**), I felt something had to be done about the now public breach. after discussion it was deemed necessary to correcting Profesor Bilgers account of events in a joint New-Zealand-German letter of reply to Science.[11] The text of this letter follows:

Ring Laser Design

"We note a letter by H. R. Bilger (3 Oct., p. 17), asserting sole authorship of the design of a ring laser C-II. This is incorrect. 1) In particular, even the inaugural design document ("The C-11 design manual," August 1994), itself skeletal in places, had two authors, Hans Bilger and Ullrich Schreiber. Early input at this level by Schreiber in the design document and from experience operating a large ring laser in New Zealand is not adequately acknowledged in Bilger's letter. 2) Bilger's letter omits mention of the more detailed and lengthy design work for C-II by the Carl Zeiss company in subsequent years. This was necessary to provide a substantial part of the novel ap-plied technology and to make the instrument possible. 3) The setup and the 'conditions of operation have changed drastically over the last 15 months. Major modifications were made during the 8-month commissioning by Schreiber on site at Christchurch, New Zealand. In summary, Bilger's letter underrates the contribution of others at all levels and ignores items 2 and 3".

No more need be said about this unfortunate incident. Rather I would gladly place on record here our (and in particlar my) immense debt to Professor Bilger in the early stages of these projects as this book should make clear. It was a great pleasure that through the generosity of a later head of department Phil Butler we were able to welcome Professors Edda and Hans Bilger back to New Zealand to see the later developmemts including UG projects Despite a period of ill health prior to his death Hans was able to lodged an article on Wikipedia documenting many considerations lying behind his design considerations for ring lasers.[12]

Both Brian Wybourne and Hans had to contend with considerable ill health, on Brians death. Hans wrote a fulsome appreciation with many memories of their time in the Southern Alps of New Zealand. And the Wild West coast of the South Island.[13] Brian's memories of Hans are similarly recorded. One often repeated story was how Hans challenged Brian to find him a student who could beat him at table tennis, when the first volunteer was demolished, but the second was a Chinese student with almost professional skills and who was victorious. Hans himself died on 13 May 2013 we at Canterbury remember the good days and are grateful for his memory. The era of large geotetic ring lasers would be unlikely to have existed without his enthusisam vision and commitment.We in New Zealand were priviledged to share in our work on C-I.

Figure 87, This historic photo is of three people of vital importance for the large ring laser program I took it appropriately in the Kepler Museum, Reigensburg

From left **1. Ritchard Falke**, from the BKG Bundesampt für cartography and Geodesy, (Ritchard had the not inconsiderable task of the necessary political effort associated with the financing of C-II and later G while working within an institution BKG which has a limited interest in promoting basic scentific research) C-II and G and their achievements are momuments to his farsighedness and success.

2. Professor Manfred Schneider of the Technical University of Munich had the vision that was vital for all the advanced projects aiming to measure flutuations in earth rotation. Schneider had already had a student build a ring laser as a pilot device for a program of geodetic observation it used argon gas as the gain medium.[14] This system had some significant defects notably the necessity for back reflection of the beam through the gain medium such defects as these were are pointed out to me in detail by Hans Bilger. He was happy to contrast these with the success of the Canterbury system for C-I as a result Schneider was convinced that the German project should be built on Canterbury lines this decision proved to be of immense importance and value for all parties. Professor Schneider recommended that Dr Ullrich Schreiber investigate ring laser measurements of earth rotation as the project for his Habilitation thesis. this thesis was eventually based on C-II.

3. Professor Ulrich Schreiber

Fig 88, Professor Ulrich K. Schreiber

Ullrich Schreiber's long and extraordinarily fruitful involvement with the projects summarized here undoubtedly make him in my eyes Professor Bilger's biggest discovery. For 20 years since his first visit Ullrich has been a regular (at least annually) visitor to Christchurch and has always making incisive contributions to the joint project. Professor Schreiber has naturally become a firm friend of the New Zealand group.

Ullrich was from the University of Gottingen he then worked with Professor Schneider of TUM and was employed at the Wetzell station to make fully operational their facility to bounce pulses of laser light from the reflectors left on the moon by the Apollo astronauts, this program has attained submilimetre precision in monitoring the earth-moon distance and monitoring its variation.

C-II was an ideal project for Ullrich's Habilitation (profesorial) thesis which took him and his family for a year in New Zealand. It was also proved a valuable stepping stone in building G.

111

Rudhi and Ulli

Figure 89, Rudhi and Ulli during a mirror change on C-II, After two accidents with C-II
it was necessary to replace the mirrors.

Rudhi from Carl Zeiss with Ulrich at Cashmere suitably suited. since they were optically contacted to the main zeodur Block. Ruhdi as an expert from Zeiss, unwrung and replaced the mirrors.

Ullrich was for a time manager of the agency Wetzell Fundamental Research Station which was operated jointly by TUM and, BKG. He still manages the G project but now following the retirement of Professor Schneider he is full Professor at TUM the Technical University of Munich.

Bob Hurst

Figure 90, Bob Hurst, His Early academic work was at University of Otago where he became New Zealand's first PhD student in astronomy (on radio telescopes), his career took him to the standards lab in PEL Wellington, then to the Department of Defence in Auckland and then joined the ring-laser project (2002 to 2014 but planning to retire July 2014)

On Clives and Morries retirement,

Graeme MacDonald

Fig 91, took over Clives (electronic) area of work, and

Rob Thirkettle

Figure 92, Rob Thirkettle (mechanical) took over Morrie.

A few other students and post docs who worked on the project

Bob Dunn

Figure 93, Bob Dunn,

once with the US Air Force and at one time involved in the Abortive Seiler Air Force Base ring laser project discussed earlier has started his own project at his institution Hendrix College, Conway, Arkansas and became a valued collaborator and welcome visitor to Canterbury before his retirement.

Geoff Stedman

Figure 94, Geoff Stedman

Now emeritus Professor of physics from the University of Canterbury in New Zealand where he gained his first degree, he then graduated with PhD from Queen Mary College University of London he did a postdoctoral fellowship there. He then returned to Canterbury as a lecturer. after spending 9 years on research in quantum mechanics as applied to the solid state and relativity he developed an interest in the wide variety of ring interferometers. partly because of the reasons discussed in chapter 4, in particular relativity. During this time he received as described earlier in this book historic letter from Professor Hans Bilger of Oklahoma State University at Stillwater. Whose effect was going to take him and the University of Canterbury down a totally new research path. His distinctions include the Fellowship and Hector medal of the Royal Society of New Zealand and the Research Medal of the University of Canterbury

Professor Brian Wybourne

Figure 95, Brian Wybourne

Brian Wybourne was always a champion of the ring laser project at Canterbury.

He had recently taken over as head of Department, his predecessor Alistair McLellan had managed the development of the Department's astronomical observatory and the building of a 1 m telescope for Mt John in the Mackenzie Basin. Fresh from this effort the mechanical workshop was in need of a new challenge and Brian saw this suggestion from Hans as a clear possibility.

Brian was himself a graduate of Canterbury and became well known for his innovative work in solid state spectroscopy[16] and his very considerably contributions to the algebraic problems arising in analysis of atomic and particle physics spectra.[17] After several spells as Head of Department at Canterbury he moved to the Institute of Physics Department ul.Grudziadzka of the Nicolaus Copernicus University in Torun Poland where he had endeared himself to the staff of the Institute and unprecedentedly although not Polish himself was later made a full professor there, the first non-Polish person to do so.[18] Several of us old friends visited him there. Brian's interests in the ring laser project continued unabated in Poland. Following several spells of ill-health of varying severity Brian died in Torun Poland on November 26 2003, We lost a good friend then.

Clive Rowe

Figure 96, Clive Rowe with Ulli

At the time Professor Wybourne was contemplating setting up the ring laser project at Christchurch a new technician had joined the electronics workshop. In mutual discussions it was agreed that he could be assigned in part to the development of this new project. Clive had taken considerable experience at the University of Canterbury as well as in industry. It was understandable that there was some reluctance in the Department to introduce a new project under a the management of a theorist. Clive and Morrie became a mainstay of the C-I enterprise. Without such hard working capable people including Morrie Poulton C-I would have been impossible. Clive brought to this an unusual set of experiences and skills.

Clives enthusiasm for amateur astronomy support of school programs etc. is legendary, he was awarded a Queen's Birthday Honours In 1999 O.N.S.M: include"Clive Henry **ROWE**, of Sheffield, Canterbury. For services to astronomy" all know Clive recognised this as a richly deserved recognition given the immense amount of help to schools and small astronomical observatories. After Geoff's severe stroke in 2001 leadership of Canterbury group operations was passed first to

Figure 97, first to Professor David Wiltshire to whom the group is most grateful for his contribution in maintaining the project And then Figure 98, on his appointment bringing considerable laser expertise to John-Paul Wells both are former Canterbury students.

Morrie Poulton

Figure 99, Morrie

For Many years Morrie Poulton has been a technician in the mechanical Workshop of the Department of Physics. Morrie had severed an apprenticeship with 'Hoffman Bearings' UK been in the Department since March 1970 and had done major work on the 1 m diameter telescope the optical craftsman telescope and Echele spectrographs for Mt John. This had given him a reputation for patience and an ability for fantastic precision with modest tools and machines ideal qualities for the ring laser project. As with Clive discussions led to Morrie being assigned the Mechanical aspects of the project for many years 1986-1999. This extended to management of the Cashmere Cavern once that became the groups principal laboratory including negotiations with external bodies such as the Cracroft Wilson family the city Council and the Historical places trust. It also extended to learning the skills of cleaning the super mirrors as necessary, and sadly the need for this was recurring and serious need. Morrie died in 2012.

Like many others in the south Island we have relied completely on David's Bell geological reports for matters relating to rock safety in the cavern. Our former student Andy Matthews was an important in liason with German scientists. Amongst the post doctoral fellows and students who worked on this project I mention Li Ziyuan, Peter Wells, Tom King, Doug Wright, Duncan McLeod, Stephen Cooper, Bryn Currie, Richard Graeme John Hold-away, Nishanthan.Rabeendran and others.

We have had a number of Prestigious visitors who include

Francis Everitt

Figure 100, Francis Everitt of the Stanford gravity probe B experiment photographed here in the cavern workshop with Richard Graham Bob Hurst and John –Paul Wells

Francis WF Everitt of the Stanford gravity probe B

Richard Packhard

Figure 101, H. Richard Packhard of the helium gyroscope (Berkeley), with UG1

Wolfgang Thierse

Figure 102, The speaker of the German parliament Wolfgang Thierse here with Professor Schreiber.

Endnotes

[1] by O.T.J. Alpers, judge of supreme court from Cheerful yesterdays Whitcombe & Tombs 1951 pp82-83
http://en.wikipedia.org/wiki/Flat_Earth_Society

[2] http://www.teara.govt.nz/en/photograph/28397/temple-of-truth-christchurch

[3] With apologies to George Gamows Mr Tompkins in paperback Cambridge University Press, 1993 pp 33

[4] physical review 1631967

[5] GEN I.

[6] Erlickson American Journal of Physics 41 1298-9 1973,
Reply to Erlickson: G. E. Stedman Am. J. Phys. 41 1300-1302 (1973)

[7] Bilger and Zavodny Phys Rev A5 591 [1972]

[8] T A Dorschner H A Hass M Holz I W Smith and H Statz Laser gyro at the on the quantum limit IEEE J Qu El QE-16 pp1376-1379 1980

[9] Brian Garner Wybourne Memories and Memoirs adam marszalek Totun(2005)

[10] Science H.R.Bilger Ring Laser design **278** 17 (3 Oct 1997)

[11] reply M. Schnieder U Schreiber and G.E. Stedman Ring Laser Design science **280** 659 (1998)

[12] http://en.wikipedia.org/wiki/Ring_lasers

[13] GEN C.

[14] B Holling G Leucks H Huder M Schenider Applied Physics B 55 46-50 1992

[15] http://www.inertialsensing.com/about-us.aspx

[16] 1965 Spectroscopic properties of Rare Earths(Wiley).

[17] http://www.mathe2.uni-bayreuth.de/axel/htmlpapers/wybourne.html
1970 Symmetry Principles and Atomic Spectroscopy (Wiley-Interscience).Wybourne, B. G. 1973: Classical groups for physicists. . Wybourne, B. G. 1998: Physics as a journey. Nicolas Copernicus University Press, Torun.

[18] [GEN C].

9 Solid state gyros

Introduction

Acoustic gyros are based on the Sagnac effect for sound waves. This has become a highly developed topic of considerable commercial interest, in which miniaturized devices whose sensor elements are fabricated from silicon crystals form the working elements of many gyro transducers for use in automobile navigation systems for example.

The comparison of counter rotating sound waves proceeds in full analogy to the optical devices described so far. Again the central physical effect is the Sagnac effect, and again these devices can be sensitive to Earth rotation which is readily detectable with small purely solid-state devices, and systems. The effect we consider here is not to be confused with the matter Sagnac effect which arises because in Quantum mechanics a moving mass has a wave associated with it with a wavelength being proportional to momentum (chapter 6). The Sagnac effect in that matter interferometer case its an extremely precise measure of the very fundamental effect because it directly reflects the mass of the moving object.

Motivation

The original motivation for this chapter was when one leader of the optical gyro industry visited and I asked him whether the principle of an optical gyro could be illustrated as a simple lecture demonstration by sound waves in a rotating ring, his answer was negative.

His reasoning for this conclusion not given to me in detail but was clearly along the lines that light was special for Relativistic arguments therefore the Sagnac effect were not applicable when the wave carrying medium is co-rotating with the rest of the instrument, the circulating speeds for the counter propagating beams could not be distinguished. I later realized that this line of argument is simply completely false. All arguments based on such standard mechanics proved valid in Newtonian science remain true in relativity. And Special relativity holds for all phenomena in any inertial frame.

Demonstrations

In fact, a perfect lecture demonstration of the Sagnac effect for counter rotating sound waves had been available for 124 years since it was already performed in 1890 in which a wine glass was rotated[1] giving audible beats. Indeed even the Sagnac effect of Earth rotation can be detected this way as a bias on the beat frequency using for example a hemispherical quartz resonator with diameter 60 mm.[2]

Earth rotation biases have also been detected and seen in the Sagnac effect in the beating of surface waves in a rotated medium[3] by Surface acoustic waves are especially clear in their resolution.[4]

Other aspects

The simple time of flight arguments used in an optical case chapter 2 can be used with equal conviction in an acoustic context also to derive the basic equations for the Sagnac effect, equally one can derive the characterize equations of the effect from relativistic arguments or from a Lagarangian analysis which allows for the Coreolis force induced by rotation[5]. This approach certainly applies to an understanding of the Sagnac effect in micro-fabricated gyro sensing chips e.g. MEMS[6] available at amazingly low cost from China now-a-days, one of our team is currently managing a project in Sweden using this technology[7]. The Coreolis force is a force which according to Newtonian mechanics comes to play when matter travels radially in a rotating medium. When two masses vibrating in plane at a frequency ω_r under a rotation at rate Ω; Due to the Coreolis effect there is an acceleration on on masses equal to $2m(\Omega \times v)$, where v is velocity. if the mode is radial, you can see how a Coreolis force can emerge which couples to a tangential mode and hence to the other radial mode modes,

and Ω the rotation rate. There is an old fable that the coreolis force should be discerned in the exit of water from bath tubs by the sense of rotation of the water, the sense of rotation supposedly being dictated by which hemisphere one is in, in fact the sense of rotation in in your bath is indictated by their geometry of the outlet and by long-lived transients motions you have created in the water. A hemisphereic corolis effect can rather be discerned in the sense of rotation of syclonic weather systems as any weather map will indicate and confirm.

Figure 103, Whilpools

One counter example to this hemispheric bath-tube fable depicted above is the clockwise [i.e. wrong for earth rotation generation] whirlpool at the hydroelectric generator intake at Lake Pukaki, South Island New Zealand the Mt Cook Range in background. However the Coreolis explanation of the sagnac effect is perfectly valid for the material gyros discussed in this Chapter.

Indeed in principle as Silverman has pointed out one can detect the effect of earth rotation in Atomic vapours. through the induced effect of optical rotation in atoms[8] Optical rotation is that of the plane of polarisation of a light beam about its direction as it passes through some medium, in this case an atomic vapour or gas where the system as a whole (atomic gas, light beam, sources and detectors rotates with the Earth. In principle, the experimental apparatus associated with Silverman's proposal need be only two pieces of Polaroid; the atmosphere would suffice for the medium. Hold your pieces of Polaroid apart, and look through both of them. Now rotate one own plane till, you get say Polaroid in its maximum transmission. Now turn on your heels, holding the apparatus steady. Do you now have to adjust the angle to get the maximum? According to Silverman, there is an optimum, non-zero angle between the orientation of the two pieces of Polaroid, at which one gets maximum transmission through both pieces. This optimum angle depends on the direction of observation, is largest when the observer looks in the direction of a celestial pole, and changes sign as one looks in the opposite direction. If (with obvious refinements of the apparatus) this effect proves to be detectable, it would mark the first demonstration of the influence of planetary motion on atomic structure.

For several decades now, alternatives to mechanical gimbals as formerly used in gyroscopes have been under development. A huge step forward was made when ring laser gyroscopes with navigational capabilities reached an advanced stage of development during the 1970s and 1980s and were given a preferred role for inertial navigation in missiles, aircraft and ships. Such navigational units typically have a triangular light beam path with an area of order 0.02 square meters. A standard aircraft gyro is sensitive enough to measure Earth rotation effects in the airplane, although this has gone unreported. Over the last two decades much larger ring laser systems for scientific application such as geophysics have been built in our laboratories. The largest such is UG2 834 m^2. See chapter 7.

This shows that the size of ring laser gyros has steadily increased over the aircraft gyros (sise of typically 60 cm^2) that had widely been taken to be the practical limit. Recently we reported a ring laser full development. The area increase directly increases the signal from the chapter 2 equation (2). The short-term precision of such devices is now below parts per billion of the Earth rotation [G^9]. Present work in this area is aimed at improving this. towards the level at which general relativistic effects like the Lense Thirring effect become measurable Fluctuations in Earth rotation induced by the moon in the form of a polar wobble are already clearly visible in UG2 chapter 7 this was also seem later in C-II data. The detecting of the Chandler wobble of the earth is visible through this technique. See entry G in chapter 7

In contrast solid-state devices we now discuss have the immense attractions of cheapness and compactness, they should not be confused with the gryo's utilizing the de Broglie wavelengths of moving particles chapter 6. Such devices bring inertial navigation into a mass market for such as rotation sensors for inertial navigation in automobiles. One strategy is to use fiber optic gyroscopes, which also are optical interferometers- see the preface, not lasers, whose area is enormously enhanced by winding the fiber in a coil, thus greatly enhancing the effective area of the device and so its sensitivity to rotation.[1] I have not myself been involved in advancing or using this particular technology though several other in the group have used it. An alterative compact technology is that of acoustic gyros. It is not easy for a physicist to keep track of the current range of the full achievement all this literature covers. For example many micro-machined silicon devices in service and under development are based not on light propagation but on elastic wave motion in solid materials. Such acoustic gyros have been neglected in the physics literature; they have in fact become increasingly useful for inertial navigation in avionics for example in a ring of silicon; one mode of vibration is coupled to another through the Coreolis force the beating of the modes gives a very accurate measure of the rotation rate. Some operate on a cantilever system:

Even the Earth's mantle suffices to show the Sagnac effect of earth rotation[10].
Helioseismology shows the sagnac effect in the acoustics waves of the sun the planets and stars for example Jupiter's Planetary waves have frequencies split by the rotation of Jupiter.[11]

This was demonstrated for example in a Chilean earthquake in 1960

in the counter-propagating tele-seismic waves in the Earth Aki & Richards[12] from a 1960 Chilean earthquake which have period 54.7 and 53.1 minutes for counter propagating waves. This means a difference in their frequencies – a sagnac splitting.

Coriolis effect in solid state gyros

Aki and Richards call this a Zeeman-like splitting rather than a Sagnac effect, referring to another parallel in physical theory in addition to those already mentioned in chapter 2. The acoustic phenomenon may be interpreted as the result of a Coreolis force in the rotating frame, the exact frequency splitting which depends on Earth structure (notably the dependence of density on radius); and.[13] Gravity affects these frequencies, for the appropriate mode in an incompressible Earth with the rigidity of steel, and so the counter propagating waves have period of 65 to 55 min.

Insects guidance systems

The Coriolis force is vital for aerobics guidance in the insect world. Many insects have appendages are known as Halteres[14] flap up and down as the wings do and operate as vibrating structure gyroscopes. Every vibrating object tends to maintain its plane of vibration if its support is rotated, a result of the Coriolis effect. If the body of the insect changes direction in flight or rotates about its axis, the vibrating halteres thus exert a force on the body. Halteres thus act as a balancing and guidance system, helping these insects to perform their fast aerobatics. In addition to providing rapid feedback to the muscles steering the wings, they also play an important role in stabilizing the head during flight. Examples are the knobs on the Crane fly behind the wings[15].

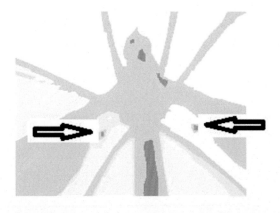

Figure 104, Halteres on a crane fly.

Endnotes

[1] Hartley Bryan, Philosophic Magazine **7** 101-110 (1890)

[2] I.V. Batov et al Izv AN SSSR Mekhanika Tverdogo Tela **27** 3-6 (1992).

[3] AM Frost, JCSethers and TL Szabo Appl Phys **48** 52-58 (1977)

[4] R.G. Newburgh P Blacksmith A .J. Burdeau and J. C. Sethares proceedings of IEEE **62** 1621-1628 (1974)

[5] GEN A.

[6] http://invensense.com/mems/gyro/itg3200.html

[7] Duncan McDonald currently leads a team on another sensor project *using solid-state gyros*.

[8] (1990) *Silverman Phys. Lett.* A **146** 175. G.E. Stedman *Physics World* November 1990 p 23

[9] GEN I.

[10] http://invensense.com/mems/gyro/documents/whitepapers/MEMSGyroComp.pdf

[11] F. A. Dahlen and M\ L\ Smith, Phil Trans Roy Soc Lond A279 44 (1995).

[12] Keiichi Aki, Paul G. Richards University Science Books, 1/01/2002

[13] p 363 of Aki & Richards

[14] the name Halteres. denotes suitable weights could enhance the performance of athletes (from ancient Greece).

10 Hidden Momentum

Nothing in this chapter has anything to do with ring lasers. Electromagnetic theory contains some puzzling paradoxes some of these appear to defy the laws of Newtonian mechanics on momentum conservation.

In standard electromagnetism it is possible to generate a boot-strap merry go round i.e. a device that can be made to rotate by minimal internal change, this is in apparent violation of the conservation of angular momentum until one takes into account the flux of momentum in the field of the (Pointing's) vector $\mathbf{E} \times \mathbf{B}/c^2$. It is also possible to make a boot-strap spaceship, namely a device that similarly can fly off into space through minimal internal change in apparent violation of the conservation of linear momentum, this problem can be traced in part to the linear momentum in the pre-existing Poynting field $\mathbf{E} \times \mathbf{B}$, this even with allowance for this momentum exchange this still violates a general theorem's in relativity which acquires the mechanical momentum by the skateboard to have pre-existed in mechanical form

To quote furry[1]. "the idea that the poynting vector is not to be taken seriously as a detailed distribution of energy flow and momentum density...Pugh and Pugh[2] conclusively vindicated the poynting vector even for the notorious case of the charged magnet where for generations textbook writers declared that their obviously could be no energy flux. (GES This timely caution not to trust every textbook on relativity must be born in mind throughout this Chapter.) this (Furry's equation 15) is the position vector of the centre of mass, This law of the centre of mass always holds for a closed system in relativity theory". This requires that apparently violates the conservation of linear momentum in relativity but can also be understood physically in terms of a hidden momentum of the charge carriers associated with their relativistic change of carry mass as their velocity carriers historically this is the way in which hidden momentum has been understood. However a change of mass with speed in relativity has become a controversial 'old-fashioned' idea for reasons I will not attempt to discuss here although used in many texts even modern research papers. The debate this introduces is not the issue at stake in this chapter. (My own research papers use the old view that I was taught but I am persuaded that the new view should be adopted, the controversy over the mass/speed discussion has been described as a conventional choice similar to the conventional choice over the speed of light discussed in chapter 4), the issue is that it is necessary to arrive at an understanding of hidden momentum which is compatible with the above mentioned theorem and with the history of the topic before one can declare that the whole matter is understood. Hidden momentum is largely unknown to physicists but I used this to try to enliven my undergraduate lectures and public lectures which supported the idea that this book with an account of some these issues (see the preface). It was amazing to me as a teacher of physics to learn that the path to a full solution of some related problems appeared only ~106 years after Maxwell's theory of (electromagnetism)[2] in which all the answers had in any case been enshrined because these equations are known to be fully compatible with Einstein's relativity. The set of circumstances I have just briefly described represents a state of affairs which is largely ignored by the physics teaching establishment. My experience has been that the debate over mass has inflamed the issue unnecessarily if I expect unavoidably. Those who deny a mass change with speed have been reluctant to provided a full mechanical explanation of hidden momentum in its historic context, so I am inclined to introduce the reader to the topics for it's general interest and as an outstanding problem in physics.

One of the great advances that Maxwell-and Faraday achieved was for concept of electromagnetic fields around each electrical charge exists an electric field **E**, and around each magnet and current carrying wire exists a magnetic field **B**. Such fields were to be thought of as little arrows spread through all space. It was also found useful to introduce other field and of special interest here is the magnetic vector potential or as it was dubbed in the 1800's 'electrotonic' field A. this concept was much discussed in the 1800's.The conventional magnetic and electric fields both depend on **A** through the equations.

$$\mathbf{B} = \nabla \times \mathbf{A}, \quad \mathbf{E} = -\nabla\phi - \frac{\partial \mathbf{A}}{\partial t},$$

The first of these equations B=∇×A says that the arrow field **A** could tell you the direction and length of the arrows comprising the **B** field imagine that the **A** field represented the rate of flow of a fluid and now dip a little paddlewheel were into it it's rotation rate. Would specify the magnitude of the magnetic field **B** and the paddle wheel axis its direction. The second equation says that its was not only charges that produced electric fields through the voltage field V changing in space but also the **A** field when it changed in time. This last effect is the basis of the electric dynamo effect discovered by Faraday. If the magnetic field within a loop of wire changes in time then an electric current would be produced in the wire. On can therefore draw a map of the vector potential by using a suitable paddle wheel, for example what I call a dipstick.

Figure 105, Dipstick for sampling **A** field

this device is a coil of wire wrapped around the edges of this large stick. If its ends are connected to an oscilloscope it will display the voltage induced in it and so the field **A**. By changes in the enclosed magnetic flux, the latter field could be generated for example by an alternating the current in a coil of wire like that below. To sample by the induced emf

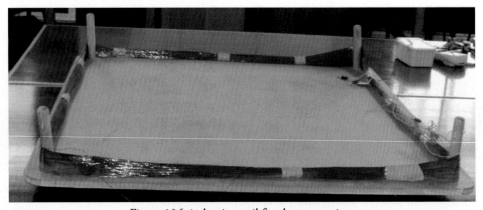

Figure 106, induction coil for demonstration

126

this coil is conveniently fed by an audio-oscilator and it is convenient to turn the large coil above into a resonant tuned circuit at audio frequencies by adding a suitable capacitor across its ends. A nice lecture demonstration is to walk over the big coil with the long dipstick to sample the A field (essentially through the bottom few inches of the long dipstick coil as the result of another equation linking **A** and a nearby current density **J**) at a variety of points through the direction and size of the maximum induced voltage **A** field $\partial A/\partial t$ the voltage induced by the. Faraday effect as the flux of magnetic field through the area of long thin dipstick coil changes in time. Near current carrying wire (at the origin below) in the coil the **A** field looks like

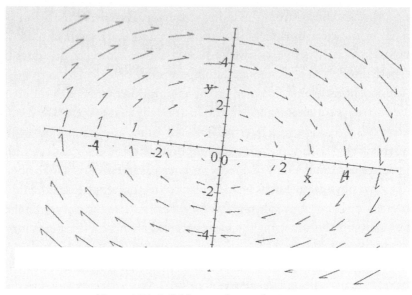

Figure 107, **A** field around a conducting wire

As I hope the reader can now imagine what one finds in the demonstration of the above photos is that a combination field A is a combination of all of these patterns from every part of the big coil, the more nearly approximating the above vector diagram when sampled near the coil wires and apparently rotating around them but fading away as one samples the field at progressively larger distances from it. I owe the basic idea of this demonstration to a very old article in teaching journal of the IOP on **A** which I am unable to trace.

With students I used to like to start the bootstrap discussion with Feynman's merry go round[3]. Consider a merry go-round which has electric charge Q on its periphery. and an axial magnetic field, produced say from a superconducting solenoid mounted axially. Now if you turn the field to zero-say by heating the solenoid through the axle, or just is the switch it off there is according to Faraday a tangential force on the charges Q∂A/∂t. This is a bootstrap merry go round in accord with the definition above. A calculation shows that the angular momentum involved is equal to the flux of momentum in the Poynting-vector field volume integral of **ExB**/c[2]. (Alternative forms of this momentum densities exist[4], for example $V\,\mathbf{J}/c^2$ where V is voltage and **J** the current density, yet another is $\rho\,\mathbf{A}/c^2$. where ρ is the density of the charge. (This form is useful for a few discrete charges, while the first form **ExB**/c[2] can be as convenient in many for computations where the symmetry is high before the fields are turned off. The magnetic force on a moving charge directly vindicates such formula looking at the impulse on a test charge moving through an **A** field. So a plain man would say momentum has been taken from the electromagnetic field and dumped into the mechanical system. (simpler and perhaps more convincing in special geometries- nice problems several American Journal of Physics articles are available point out that the use of cylindrical symmetry often makes the mathematics readily soluble.[5])

Having looked at a bootstrap merry go round now look at a book strap spaceship. Put + - charges at opposite ends of a skateboard wheels turned 90 degrees to its axis. Now add a solenoid as before to give a magnetic field through the skate-board. Turn off the current and Faraday induction produces an **A** field at each charge which generates a forces driving the skateboard in a direction perpendicular to its the long axis. Off you go into space. But you were told in early physics classes that momentum was conserved so where did that momentum come from? But to find where it could have come from you take the system apart and there seems nothing. Again the combination **E×B** that existed before the fields were turned off was a linear field with a flux over space and when you calculate it indeed it comes out right. And indeed the point of the Furry theorem quoted earlier[6] is that the center of mass of a closed system is un-accelerated in special relativity; the obvious solution to the puzzle then is that momentum which has been transferred from the electromagnetic field to the skateboard through the impulse on the skateboard from the induced voltage. Detailed calculation show corresponding to a flow of momentum of the electromagnetic Poynting flux and skateboard impulse are indeed of equal magnitude. This conclusion exactly fits Furry's work quoted above. However this does not solve the entire puzzle. Not only do these mome`nta of skateboard and field prove on calculation to be equal and opposite, but suppose that the hidden momentum calculated on the assumption that mass of the charge carriers changes with their speed as the current is turned on and off, and historically that was the way hidden momentum was understood as my references will confirm(I refer to Penfield, James and Furry) just as it is now unpopular today to talk about mass varying with speed so it was unpopular in the early days of hidden momentum to think of the Poynting vector as serious a flow of energy and mass as in Furrys remarks quoted above. Students enjoy trying to identify what physical mass was moving prior to the current being off, but inevitably fail, Both the solutions to the mystery out lines above are highly non-intuitive. The curiosity still remaining is that if one does a calculation of hidden momentum using the idea that mass changes with speed, it turns out to have the same magnitude as the momentum flux in the pointing vector, or the skateboard impulse. And historically this is what was called 'hidden momentum' unknown until the 1950's although buried in Maxwell's equations.[7] My Penfield and Haus references (reflects their adherence to the old view that mass varies with speed as is clear in their equation 5.22, and its application to what is essentially my bootstrap spaceship as stated in their figure 7.9).[8] This is clearly is the monograph alluded to in the last reference of the (Shockley and James)[9] this paper has clearly imbibed the old idea, they represent their current carriers using two rotating disks with oppositely charged peripheries. I find that this latest calculation of hidden momentum can be transparently done by following up this idea, and adding a constant like c to all carrier speeds, and indeed one can imagine a limiting case in which the charge carriers approach the speed of light in frictionless tubes. With negative and positive charge carriers being confined to separate tubes a little bit like the two lasing beams in a ring laser (despite the disclaimer at the head of this chapter, not unlike ring lasers except that + - carriers cannot follow the same paths if they are to avoid mutual annihilation). They carriers accelerate and decelerate under the electric field from the point charges, pressing on the tube that confines them this admittedly bizarre model is nevertheless calculable. As they do so the resulting change of speed in the charge carriers go faster at one end of their circuit and so have more mass on the old fashioned view and so more momentum and when you calculate this effect, the speed affects the mass of the charge in the correct way to gives an extra momentum flux hidden in the carriers and in exactly the right sense and magnitude to match the momentum flux in the Poynting vector. The proof of the equality of these quantities is not trivial but possible but it is necessary to incorporation of the full history of the acceleration of the charge carriers under the electric field from its first creation, once this is done for the Bootstrap space-ship, hidden momentum in the historic sense is neatly exchanged in totality with the field momentum. Haus[10] illustrates this historic understanding of hidden momentum neatly using an Escher drawing 'Encounter'

If this historic view of mass variation with speed is as wrong as I am told and I accept and that the relativistic expression for particle momentum, $\mathbf{p}=\gamma m_0\mathbf{v}$. All I can say is that I do not know of anyone who has attempted an explanation for this equality I describe above based on the Panfield-Haus model of the result of a calculation on the more approved one. I have recently found one account at a calculation[11] which claims a formulation of hidden momentum integrated with that of the Poynting flux and proves their equality, this in agreement with Furry their identification of hidden momentum is obscure to me. So in such an account the hidden momentum seems not to perceived as essentially relativistic in the manner of Penfield and Haus, to my amazement even this is now(1861-2009) 148 years after Maxwell. The story for hidden angular momentum as opposed to linear momentum (which was the relevance to Furry's Centre of mass theorem) and which is obviously relevant to could also do with clarification bootstrap merry go round. Has only been briefly discussed[12] see also Furry. For these reasons I do not consider the story of hidden momentum in special relativity a closed one.

A final thought in conclusion

I think I can fairly claim that in this book I have proved my commitment to my belief that all physics is based on experiment. But in relation to that 148 year gap I referred to above the theorist in me wonders how many other such unsolved physics problems within currently accepted theories will still be quietly waiting to be solution after the end of the next century[13].

Endnotes

[1]   Furry American Journal of Physics 37 621-636(1969)

[2]   E.M. Pugh & G.E. Pugh Amer J. Phys. 35 153 (1967)

[3]   106=(1861=Maxwell's theory (http://en.wikipedia.org/wiki/History_of_Maxwell's_equations) -1967=Shockley and James Physical Review Letter, referenced below)

[4]   Feynman R P, Leighton R B and Sands M 1965 The Feynman Lectures on Physics vol 2 (Reading, MA: Addison-Wesley) pp 17-4–17-5, 27-11 .

[5]   Calkin M G 1966 Linear momentum of quasi-static electromagnetic fields Am. J. Phys. 34 921–5

6   Romer, R. H. American Journal of Physics, Volume 34, Issue 9, pp. 772-778 (1966). also see the calkin reference above.

7   Furry's American Journal of Physics 37 621-636(1969)

8   Robinson Phys Reports 16 313-354

9   Penfield and Haus Electrodynamics of moving media Technology Press Cambridge Massachusetts 1967 p 125

10   Shockley and James 18 876-879 (1967)

11   H A Haus Electrodynamics of Moving Media and the Force on a current Loop, applied Physics A A27 99-105 (1982)

12   D. Babson S.P. Reynolds R. Bjorkquist,.and D.J. Griffiths American Journal of Physics 77 826-833. (2009)

13   G.E. Stedman Physics Letters 81 320 (1973).

14   the situation is reminiscent of my experience in finding one particularly long standing error in the work of Feynman. G. E. Stedman Diagram Techniques in Group Theory (Cambridge University Press1990, 2009 p 279,

15   see Wikipedia crane fly.

11 General References

General and reviews

> A. G. E. Stedman *Ring laser tests of fundamental physics and geophysics*, Reports Progr. Phys. **60** 615-688 (1997).

>B. R Anderson, H R Bilger and G E Stedman, *"Sagnac" effect: a century of earth-rotated interferometers,* Am. J. Phys. **62**, 975--985 (1994).

>C Lydia Smentic: Brian Garner Wybourne Memories and Memoirs adam marszalek Totun (2005)

>D. An elementary proof of the geometrical dependence of the sagnac effect D. R B Hurst,. J-P R Wells and G E Stedman *J. Opt. A: Pure Appl. Opt.* **9** 838–841, (2007)

>E. Rizzi, G., Ruggiero, M.L. (eds.): Relativity in rotating frames Kluwer Academic Publishers, Dordrecht, 2004

>F. G. E. Stedman Ring Interferometric tests of classical and quantum gravity *Contemporary Physics* **21** 311 (1985)

>G. R. Anderson, I. Vetharaniam and G. E. Stedman, Conventionality of synchronisation, gauge dependence and test theories of relativity *Phys. Reports* 295 93-180 (1998)

>H. *Physical Relativity* by Harvey Brown Oxford University Press 2005

>I. Karl Ulrich Schreiber and John-Paul Wells 'Large Ring Lasers for rotation sensing' *Reviews of scientific Instruments.***84** 041101 (2013)

12 INDEX

R

Raytheon 107-8
relativity 3-5, 12-16, 41-7, 50-5, 57-8, 60-3, 107, 109, 114, 119, 125, 128, 131
Rhuddi 78
Richards 122
Risks of success 71
Rizzi 42, 45, 131
Ron 5, 41-2
Rowe 33-4, 69, 74, 95, 99, 106, 115
Rutherford 63

S

Sagnac effect 1, 3, 7, 11-15, 22, 45-9, 59, 75, 119-22, 131
satellites 73
Schneider 76, 94, 96, 111-12
Seiler 108-9, 113
shot noise 71
Silverman 34, 120
simultaneity 3, 42, 46, 50
Single mode 18, 71, 75, 86, 109
slow clock transport 44, 50, 55
sound waves 11, 119
Spectra 62-3, 109, 114
STROBES 74-5
super-mirrors 17-18, 24, 73
synchronisation 14, 44-5, 47-8, 57-8, 131

T

television 31, 34, 39, 86
Tendentious 13
time dilation 44, 50, 55, 57-8
Torun 94-5, 114-15, 118
Tucker 14-15, 48
TUM 112
twin paradox 57-8

U

UG1 21, 38, 69-70, 72, 78, 81, 86-8, 101-2, 117

UG2 5, 39, 69-70, 86-8, 102, 121
UMU 82, 99
University of California 98

V

vacuum 12, 53, 74, 77
ventilation shaft 28, 31-2, 34
Vuky 15

W

wavelength 11-13, 16-18, 21-2, 62, 67-8, 73, 77, 102, 108, 119
Weber 77, 80, 94
Weinberger 48
Wetzell 5, 36, 77-8, 111-12
Weyl's 46
Will 4-6, 11, 13, 17, 23, 46, 48-52, 55-9, 61-2, 65-8, 73, 79, 86-7, 95-6, 99-101
Wilson 28, 31-2, 116
Wiltshire 115
Wybourne 21, 24, 33, 94-5, 106, 109-10, 114-15, 118, 131

Z

Zerodur 16-17, 21, 71-2, 75-8, 90, 95, 108
Ziyuan 95, 113

Printed in the United States
By Bookmasters